EXPLOSIVE LOADING OF
ENGINEERING STRUCTURES

Explosive Loading of Engineering Structures

A history of research and a review of recent developments

P. S. Bulson

Consultant to the Mott MacDonald Group
Visiting Professor of Civil Engineering, University of Southampton

CRC Press
Taylor & Francis Group
Boca Raton London New York

CRC Press is an imprint of the
Taylor & Francis Group, an **informa** business

A TAYLOR & FRANCIS BOOK

CRC Press
Taylor & Francis Group
6000 Broken Sound Parkway NW, Suite 300
Boca Raton, FL 33487-2742

First issued in paperback 2019

© 1997 Philip Bulson
CRC Press is an imprint of Taylor & Francis Group, an Informa business

No claim to original U.S. Government works

Typeset in 10/12 Times by Blackpool Typesetting Services Limited, UK

ISBN-13: 978-0-419-16930-7 (hbk)
ISBN-13: 978-0-367-86634-1 (pbk)

A catalogue record for this book is available from the British Library

Publisher's Note
The publisher has gone to great lengths to ensure the quality of this reprint but points out that some imperfections in the original may be apparent

Visit the Taylor & Francis Web site at
http://www.taylorandfrancis.com

and the CRC Press Web site at
http://www.crcpress.com

Contents

Contents

Preface

A long time ago I walked over a snow-covered marsh to place explosives near unsafe grenades; later cutting charges were set on steel joists under a desert sun, and in another part of the world camouflets were bored in the rock of monsoon-lashed mountains. Tables were used in military handbooks to calculate charge weights, and instructions were read from the same books on the safety of handling and storage. We never asked who calculated the figures or drew up the precautions, or where the history of the subject could be found. We were told to concentrate, keep cool and make sure that we avoided blowing up the explosive store, the headquarters bunker and above all, the Colonel. The source or level of accuracy of our technical information was not questioned by us.

The world of explosive technology and the science of explosive loading has come a long way since those days, but it was always a secret world. The development of military weapons, the response of civil and military structures, and the legal overtones of great disasters involving dust, vapour and gas explosions all require circumspect behaviour by those in the know. Scientists and engineers conduct target response tests in nether regions or behind high fences. Analysis and simulation are hidden in a parallel world of security passes and personal vetting. Much research is subject to restrictions on open publication, and the history of research and development is not always available to newcomers in the field. Nevertheless, whether we are concerned with gas, vapour, dust, plastic explosive, Semtex or nuclear warheads, the need to be aware of history and to use a sense of history in judging the way forward is still desirable.

This book is an attempt to review developments over the years in methods of calculation, measurement and prediction of the dynamic loading on structures from explosions. It endeavours to trace the history of the subject and to summarize some of the latest published thinking, and considers finally a range of structures from buildings and bridges to ships and aircraft. It is not concerned with the design of protective structures or the hardening of existing structures, but there are brief passages on the analysis of response, residual strength and safety. The work and achievements of major contributors to the subject over the past century are discussed, because it is important to realize that the subject has been taken forward by a collection of mathematicians, physicists, chemists, military researchers and manufacturing experts as well as by engineering scientists specializing in the interaction of loads and structures.

The fundamental science of explosive loading has been relatively slow to develop because the level of independent academic investigation of the subject in universities and similar institutions has been less than some would have wished. Apart from the cost of facilities and the difficulty of ensuring safety, many authorities supporting research with a military flavour have been the subject of attention from pacifist groups. These groups, of course, never have to grasp the nettle of responsibility for the protection of our population, our heritage and our way of life from destruction by explosion or threat by terrorism. It is therefore hoped that this survey will help to increase interest in the subject by the academic workers in structural behaviour.

In the preparation of this book, I have received much administrative help from the Mott MacDonald Group, and benefited from continuing contacts with the Defence Research Agency and one or two universities. I am indebted to two word-processor experts, Miss Rosemary Mardles and Mrs Sue Starks, who have given legibility to the script, and to the editorial staff of Chapman & Hall, who waited patiently for the final draft.

P. S. Bulson
Winchester
1997

Note: Because this is a work with many references to history, readers are asked to accept a mixture of fps and SI units in the text. A conversion table is provided at the end of the notation section on p. xv.

Acknowledgements

The reports from which the references to research in support of UK activities during the Second World War have been drawn (e.g. the publications of Christopherson and others) are available from the Institution of Civil Engineers Library, London, or from the UK Public Records Office at Kew. A paper by Walley [8.20] gives full details of these collections.

The major part of section 8.2, dealing with the effects of explosions on civil bridges, has been taken from work carried out by the author with the support of the Defence Research agency (Chertsey).

The review of historical developments in explosive technology given in the Introduction and Chapter 1 has been drawn from a number of sources dealing with scientific history, including textbooks, the Encyclopaedia Britannica, and the proceedings of learned societies.

The cover painting is by the author.

Notation

All symbols are defined in the text where they first appear. The symbols listed below are those that appear repeatedly, or are of greatest interest.

LOWER CASE LETTERS

a	side dimension of an explosive charge
	a constant
a_m	maximum bubble radius
b	side dimension of an explosive charge
	a constant
c	seismic velocity
	velocity of sound in water
	maximum aggregate size
	fragmentation parameter
	viscous damping value
c_0, c_3	speeds of sound for gasses in chambers
c_1, c_2	coefficients
c_d	drag coefficient
c_s	seismic wave velocity
d	water depth
	depth of explosion
	distance from mouth of shock tube
	projectile diameter
f	depth factor, ground shock coupling factor
	hydraulic coefficient of friction
h	thickness of an explosive charge
	height of roughness in a pipe
i	specific impulse
k	spring stiffness
	soil characteristic
k_p	pressure factor in tunnels
k_i	impulse factor in tunnels

l	clear span between supports
	length of explosive charge
m	mass of projectile
m_e	equivalent mass
n	attenuation coefficient
	circumferential modal index
p	pressure, overpressure, stress in soil, vertical penetration
p_a	atmospheric pressure
p_d	drag pressure
p_e	duct entrance static pressure
p_i	initial pressure, peak stress in soil
p_m	peak pressure
p_0	peak overpressure, static yield pressure
p_r	reflected pressure
p_s	pressure outside duct
p_1	instantaneous pressure, peak instantaneous pressure
	pressure at tunnel entrance
p_{max}	maximum pressure
p_x	peak pressure at a distance x along a tunnel
p_w	internal overpressure in a shock tube
q	dynamic pressure due to blast winds
q_L	equivalent uniformly distributed load
q_D	dead load per metre
r	pipe radius
	radial distance
s	distance
	charge spacing
t	time
	slab thickness; thickness of casing
	time measured from moment of arrival of blast wave
t_a	time of arrival of shock
t_A	intervals
t_c	time of reflection effects
t_d	duration of blast pressure
t_p	minimum slab thickness
t_m	time of maximum response
t_0	time for shock front to reach given point
t_r	duration; rise time
u	dilational seismic velocity
$\bar{u}$	shockwave velocity
u_a	speed of sound in air
u_p	permanent horizontal movement
u_x	outburst speed
υ	shock front velocity

v_p	permanent vertical movement
v_c	seismic velocity in concrete
v_s	speed of surface of a charge
w_u	maximum permissible displacement
x	surface distance
y	depth of a point below the surface
y_m	shock strength; maximum deflection
y_{el}	deflection at the limit of elastic behaviour
y_{st}	static deflection
z	depth

CAPITAL LETTERS

A	area; cross-sectional area
	positive impulse of an explosion
D	crater diameter; cylinder diameter
	fireball diameter
	duct diameter; tunnel diameter
E	instantaneous energy release
	energy factor
E_p	perforation energy
$\bar{E}$	dimensionless decay parameter
F	soil factor
	resisting force; nose force
F_e	equivalent peak force
H	depth of burial
	surface wave amplitude
	height of charge
	target thickness
I	positive impulse per unit area
I_0	peak impulse
K	constant in penetration formula
	mass scaling factor
	constraint factor
K_L	transformation factor
K_M	transformation factor
K_R	transformation factor
L	length
	distance along a tunnel
	track length
	span of beam
M	virtual mass
M_A	fragment weight distribution parameter
M_L	loading modulus

M_U	unloading modulus
M_S	Mach number
N	nose performance coefficient
	number of fragments
Q	work done during expansion of explosive gases
R	radial distance from the centre of an explosion
	distance above ground
R_e	effective slant range
R_{me}	maximum resistance
S	unsupported span
	soil constant
	cross-sectional area of a tunnel
T	time of penetration
	thickness of slab
	decay time constant
T_1	period of oscillation
T_2	period of oscillation
T_e	static temperature
T_s	static temperature
T_{te}	stagnation temperature
T_{ts}	stagnation temperature
V	impact velocity
	chamber volume; volume
V_c	critical velocity
V_L	propagation velocity
V_s	striking velocity
W	weight
	yield of explosion in megatons
W_c	casing weight
W_f	fragment weight
W_l	mass per unit length
W_p	total projectile weight
X	transverse distance
Y	radial flow of water

GREEK LETTERS

α	peak stress attenuation factor
	angle between a radial line and the vertical
	time constant
	factor applied to distance to establish peak pressure at tunnel entrances
β	time constant
γ	ratio of specific heats

γ	ratio of specific heats
γ_0	gas constant for expansion chamber
γ_3	gas constant for compression chamber
θ	angle of orientation of a bomb
	time for pressure to fall to half initial value
μ	refractive index
ρ	mass density of soil
ρ_a	atmospheric density
ρ_0	density
σ_a	flow stress
σ_c	crushing strength of concrete
σ_r	reflected stress
σ_u	ultimate tensile strength
σ_z	peak vertical stress
2ϕ	cone angle

Conversion factors

Property	SI units	fps units
Length	1.0 mm	0.0394 in
	1.0 m	3.28 ft
Area	1.0 mm^2	0.00155 in^2
	1.0 m^2	10.76 ft^2
Mass	1.0 kg	2.205 lb
Density	1.0 kg m^{-3}	0.0624 lb ft^{-3}
Force	1.0 N	0.255 lb f
	1.0 kN	225 lb f
Stress	1.0 kN m^{-2}	0.145 lb in^{-2}
	15.44 N mm^{-2}	1.0 ton f in^{-2}

Introduction

Explosions can threaten people's lives. They can also threaten the integrity of dwellings, industry and the security of communications, transport and services. Explosions can be man-made or result from tragic accidents, and can range from nuclear explosions to the firing of a shotgun; or from the detonation of unconfined vapour clouds to a bursting tyre. Explosions can be used as weapons of war as well as instruments of peace. The military sappers and miners of one generation become the quarry blasters of another.

Almost every evening our television screens show the effects of explosions on structures. A shattered hotel, a damaged police post, a domestic gas explosion, the explosive failure of an aircraft pressure bulkhead or of a jet engine. In spite of this the engineering profession in general is not well versed in the design of static or moving structures to withstand explosions, partly because in the past specifications have rarely included explosive loading as a factor in design, and partly because the various dynamic effects of explosions on structures have only been examined as research subjects in a small number of research laboratories.

The cost of providing a safe environment for research and testing is high, and experimental work in most countries has been left to the armed services, government research establishments, or to large industrial explosive manufacturers. Very often the results are not openly reported because of security restrictions. The author, however, senses a change, and a growing need for information. The risk of chemical plant explosions is a demanding design problem. A level of protection against nuclear and conventional weapons of attack is often specified in new civil works, and there is a terrorist danger, which threatens marine and ships' structures as well as buildings, bridges and aircraft. The Gulf War in 1991 emphasized the structural damage that can result from modern weapons of great accuracy.

Where can the designer turn for advice? There are several good texts on the physics of explosions, the science of detonics and the design of protective structures. A fund of knowledge lies in the archives of government departments and the services on the testing of devices of war. The collection of wartime home security reports in the British Public Records Office is worth travelling a long way to read and a recent paper has summarized this work. The output of laboratories and agencies in the USA since the Second World War has been of great value and many important technical memoranda have been published over the years by the American military services. There have been many conferences recently on related subjects, but their proceedings have not always been widely distributed. Military and Naval research has been aimed at the effects of explosive attack weapons, and at the improvement of defences against them, but the results are closely guarded. Much of the work has been *ad hoc* testing in support of development, with no specific examination of the fundamentals, and without the discovery of parameters and coefficients that could influence future design and research.

This book has been written to try to collect the historical philosophy of the subject together in a way that will be a useful introduction to scientifically minded newcomers and a quick reminder of the fundamentals to experts in the field. Where possible the principles have been drawn out to reveal the underlying science. Where the principles are clouded, an attempt has been made to clarify them by examining other methods of approach. It seemed logical to begin with a review of the nature of explosions, so we deal first with the basic science, starting with a brief history. The beginnings of gunpowder, or black powder as it was called, go back a very long way, and volcanoes and natural gases have been exploding since the morning of time; but most modern work stems from the discovery of nitroglycerine by the Italian scientist Sobrero less than 150 years ago. His name is less well known than that of the two Nobels, father and son, who invented the simple manufacturing process and discovered the importance of shock, rather than heating, as an initiator.

The physics of the detonation process is of interest, from initiation to the formation of the shock front and blast wave. This has been well documented in works by Kinney and Graham, Baker, and Henrych. It must be remembered, too, that explosions can take place underwater and underground, as well as in air, and the influence of the surrounding medium can be considerable. Many man-made explosions, particularly in weapons of war, are accompanied by the fragmentation of a disintegrating casing. Nuclear explosions are followed by hurricane-like winds of great magnitude. Structures in the explosive field can be damaged by the instantaneous rise of air-blast pressure, by fragments, and by blast winds. In nuclear explosions humans can be killed by radiation effects, although this aspect is not a subject for the book.

It was not until 1919 that the scaling laws for simple explosions were expressed succinctly. A presentation by Hopkinson to the British Ordnance Board was not conveyed mathematically, but was of classic significance. He pointed out that if two structures were made to the same drawings and of similar materials, but on different scales, and if charges were detonated to produce similar structural effects, then the weights of the charges needed to be proportional to the cubes of the linear dimensions. This law was apparently discovered independently in 1926 by Cranz.

Bertram Hopkinson was an outstanding researcher. His father, John Hopkinson, was a consulting engineer and inventor who had applied his scientific skill to the development of the quality of optical glass, and to its use for lighthouse illumination. Bertram was born in 1874, was a day boy at St Paul's School, and had won a major scholarship to Cambridge before he was 17. He read for the Mathematical Tripos. When he left university at the age of 22 he entered the legal profession and was called to the Bar in 1897. He was on his way to Australia to carry out a legal inquiry when he received news of his father's death in a climbing accident in Switzerland. Bertram was recalled to London, and immediately decided to give up his legal career in order to complete his father's unfinished engineering work. With his uncle, Charles Hopkinson, and assisted by Mr Talbot, he was subsequently responsible for the design of electric tramways at Crewe, Newcastle and Leeds. The partners wrote a paper on Electric Tramways for the Institution of Civil Engineers in 1902, and were awarded the Watt Gold Medal. In 1903, at the age of 29, Hopkinson was elected to the chair of Mechanism and Applied Mechanics at Cambridge. For the next ten years, under his leadership, the Engineering School increased in numbers and repute.

He served as joint Secretary of the British Association Committee on Gaseous Explosions, whose reports contain records of many of his experiments. At the outbreak of the First World War he dropped all other interests and obtained a commission in the Royal Engineers. He concerned himself with the collection of intelligence, and in conducting experiments in connection with an arrangement for the protection of warships from the effects of mines and torpedoes, the hull 'blister'. The blister was able to absorb the energy of an explosion by becoming deformed. Just before the war he had written papers on the measurement of pressure due to the detonation of high explosives or by the impact of bullets (the famous 'Hopkinson Bar' experiments), and on the effects of the detonation of gun cotton. He thus became the foremost expert in Britain on the protection of ships and structures against explosions, and also on the design of attack weapons, such as aircraft bombs. From this he became interested in the armament of military aircraft generally, and worked at a secret experimental station at Orfordness. By June 1918 he had become Deputy Controller of the Technical Department of the Air Force. He learned to be a pilot and generally flew alone. On 26 August 1918, he crashed near London in bad weather and was killed.

Bertram Hopkinson was aloof from any consideration of private advantage or personal convenience. Nothing ruffled his serenity or impaired his judgement. His death at the age of 44 was a tragedy and the science of flame and explosion suffered a great loss. This summary of his life and work has been taken from a Royal Society memoir. His scientific papers were collected and arranged by Ewing and Larmor in 1921 and published by the Cambridge University Press.

I have dwelt at length on the history of Hopkinson for two reasons. The first is to underline that Hopkinson, because he was not trained within the high walls of a narrow engineering field, was free to apply his creative thought across the whole area of Applied Mechanics, Physics and Electrical Engineering. This breadth of experience led to the discovery of the explosive scaling law which was mentioned earlier and which is still such a fundamental part of our analysis of the effects of explosions on structures. It underlines the need for cross linking in our training of engineers and scientists. The second reason is to emphasize the quality of the British contribution to the science of explosions in the early part of this century.

The fundamentals also owe much to three other outstanding British scientists. The first of these was Horace Lamb, a mathematician born at Stockport in 1849 who was professor of mathematics in Manchester University from 1885 to 1920. He was the recognized authority on hydrodynamics and wave propagation, among many accomplishments, and it was he who set down in his famous book *Hydrodynamics* the physics of plane waves diffracted by striking discs, cylinders or circular apertures in plane screens. He also focused attention on the work of Riemman on the formation of shock waves, and to the earlier work of Earnshaw on the mathematical theory of sound.

The second scientist was William John Macquorn Rankine, born in Edinburgh in 1820, who was trained as an engineer. He became professor of civil engineering at Glasgow University in 1855, and died at the relatively early age of 52. He was the author of the first formal treatise on thermodynamics, as well as a remarkable manual of civil engineering. He discovered the changes in pressure, density and velocity of a gas passing through a shock wave, and published this work in 1870. This analysis was of great significance in the study of the behaviour of explosions, and was also discovered by Hugoniot in Paris in 1889. Rankine, like Hopkinson, was outstanding in a number of areas of physics and engineering at the same time.

The third scientist was Geoffrey Ingram Taylor whose work on the dynamics of blast waves from explosive charges was of great value to the defence research effort in Britain in the period between 1936 and 1950. His earlier papers dealt with the propagation and decay of blast waves from conventional weapons, but his later work was devoted to the behaviour of blast waves from the first atomic explosion in New Mexico in 1945, and from underwater atomic explosions. Readers are recommended to examine the collection of the scientific papers of Sir Geoffrey Ingram Taylor, edited by G. K. Batchelor and published by the Cambridge University Press in 1963. Volume 3 of this work deals with the aerodynamics and mechanics of projectiles and explosions.

We are concerned in this book with the shock and dynamic loading acting on structures from various types of explosion. The accuracy of the figures depends on the quality of the instrumentation used to measure instantaneous pressure, the variation of pressure with time, the duration of the pulse, and the velocity of associated impulsive effects such as blast winds. In certain instances the response of the structure influences the nature and level of the applied dynamic loading.

Accurate measurement of pressure and duration first became historically important during the rapid development of the science of military ordnance in the seventeenth and eighteenth centuries. It was important to know the pressure that resulted from the firing of gunpowder, which was needed to estimate the required structural strength of gun barrels, and to calculate initial projectile speeds. The appearance of gunpowder and artillery in Europe started in the thirteenth century, and Berthold Schwarz, a German, is usually credited with the honour of harnessing the propellant force of gunpowder. Roger Bacon gave an account of the composition of gunpowder in the middle of the thirteenth century, and artillery in the form of cannons seems to have been introduced to the battlefield in the fourteenth century. The development of the material and manufacturing methods for cannons proceeded steadily via cast bronze or brass (gunmetal) to wrought iron (longitudinal rods and rings) and cast iron by the sixteenth century.

In England, the underrated Benjamin Robins wrote his *New Principles of Gunnery* in 1742, in which the first chapter dealt with force of gunpowder and the pressure on a ball at the instant of explosion of the powder. The pressure was needed to calculate the velocity of the ball when leaving the muzzle of the cannon. Robins, whose scientific work in this field has been admirably summarized in recent times by Professor W. Johnson in a paper in Volume 4 of the *International Journal of Impact Engineering*, was dealing mainly with relatively lightweight cannon balls weighing about twenty-four pounds. To find the force on them he describes the burning of gunpowder at the top end of a closed tube almost filled with water. He found that an ounce of gunpowder expands to a volume 244 times as great at room temperature, and to a volume 1000 times as great at the temperature of 'red hot iron'. This told him that the initial pressure on the shot would be about 1000 atmospheres and he used this to calculate the muzzle velocity of the ball.

Robins also introduced the concept of the ballistic pendulum to measure projectile velocity. This consisted of a block of wood attached to an iron pendulum, which was suspended from a metal crosspiece supported by a substantial three-legged frame. The movement of the block when struck by a projectile was recorded by means of a ribbon attached to the underside of the block, and passing over a transverse member attached to the supporting frame. Robins, who was a mathematician, also proposed that large guns should have rifled barrels, but a century was to pass before this idea was taken up.

Ballistic pendulums were increased in size and strength during the eighteenth century but it was not until the latter half of the nineteenth century that major new developments occurred. Firstly, Alfred Krupp in Germany produced an all-steel gun, drilled out from a single block of cast metal, and then Captain T. J. Rodman of the Ordnance Department of the US Army produced a gauge for the measurement of the pressure exerted by the burning of gunpowder in a closed vessel. This was a useful development, described in Rodman's fifth and sixth reports on his experimental work, and was in the form of an indentation apparatus, known as a cutter gauge. A gun was prepared by drilling transversely at intervals so that a plug containing a cylindrical indenting tool could be inserted. When the instantaneous pressure acted on the end of the tool, it drove the other end (which was shaped as an indenting point) into a copper disc and the amount of penetration was an indication of the pressure. Apparently this method was originally proposed by a Major Wade and it enabled the effect of increasing the weight of the charge and the weight of the projectile to be assessed accurately for the first time. As might be expected, the peak pressure due to the explosion was found to diminish rapidly as the distance along the gun barrel from the source of the explosion increased.

During the first part of the twentieth century, particularly during the years leading up to both world wars, there was a large increase in the amount of thought given to methods of measurement. Other mechanical gauges that were developed by organizations such as the US Naval Ordnance Laboratory or the UK research department at Woolwich Arsenal included the ball crusher gauge, in which a spherical copper ball was compressed between a sliding piston, activated by the pressure, and a fixed anvil. The permanent deformation of the ball gave an indication of pressure. A number of varieties of spring-piston gauge were also developed, and several types of foil and cylinder gauge were used in practical experiments. The foil gauges consisted of thin diaphragms of copper or aluminium foil, or even paper. These were clamped over the open ends of cylinders connected face-on or side-on to the blast, and the deformation of the discs gave a measurement of pressure. Other gauges used aluminium strips or wires clamped as cantilevers to steel posts, to form cantilever 'flags'. The permanent tip deflection of the cantilevers varied with charge distance for various charge sizes. Simple gauges of this type were often used after the Second World War when there was great research activity into the nature of nuclear explosions. The ultimate in simplicity was the deployment by the British scientist, William Penney (later Lord Penney), of petrol tins at the atomic weapon test on the Bikini Atoll. The degree of deformation of the tins by the blast wave, in terms of the change in internal volume, enabled a good approximation to be made of the magnitude and distribution of the peak pressures. These test measurements often succeeded when more sophisticated gauges failed. Later this technique was used in experimental research on conventional explosions.

The greatest step forward in pressure measurement, however, was the development of the pressure transducer, using piezoelectrically active crystalline materials that produce electric charges when strained. Tourmaline, quartz and ammonium dihydrogen phosphate are substances that have been used in successful gauge design, which in addition to the crystals require amplifiers, calibration circuits and electrical recording systems. Condenser-microphone gauges have also been developed, using changes in capacity under pressure, and resistor gauges which use changes in resistance under stress as the measurement. All of these gauges are particularly useful for specified characteristics such as low-frequency response, pressure range, duration and temperature. Many defence laboratories and research institutes in the UK and USA have produced their own versions of the pressure transducer, ranging in size from miniatures with diameters of a few millimetres to large-scale components with diameters of 10 centimetres and more. These and others have been reviewed in detail by Wilfred Baker in his book *Explosions in Air*, published by the University of Texas Press in 1973 in conjunction with the Southwest Research Institute of the USA. In a most useful survey he also discussed the design of gauges to measure dynamic pressures due to blast winds and to record the time of arrival or time of start of release of blast energy.

The accurate measurement of the characteristics of the dynamic loading due to explosive shock has, as we have seen, been of great interest to structural engineers. The instrumentation for a major nuclear or high explosive test can be astronomically expensive, but unless the pressure/duration characteristics of a blast wave at various distances from the source and at various parts of a structure can be measured the calculation of structural response in air, water or underground cannot be carried out with scientific confidence. Care must also be taken to distinguish between detonation and deflagration, the absence of sudden shock, and changes in the shape of the pressure/duration curve. These factors are pursued later.

It is important in the analysis of loading and response not to overlook the lack of uniformity in explosive actions. The pressure distribution depends on the shape of the charge, but even with a perfectly spherical charge under ideal conditions it would be wrong to assume that the peak pressure at a given distance from the centre of the sphere was uniform and regular. There are peaks and troughs in the distribution pattern that make it difficult to treat the loading analysis as an exact science. Charges of nominally the same weight and geometry do not necessarily yield similar pressures, durations or impulsive characteristics. The physical conditions are so variable that great care must be taken when drawing analytical conclusions, because experimental scatter in laboratory or field tests is high.

Some of the earliest methods of calculation of structural response were associated with the assessment of the strength and safety of gun barrels. Cast guns often burst on discharge because of minute flaws, and in the fifteenth century in England accidental bursts of military ordnance were considered important enough to alert the privy council. The idea of proving the strength of ordnance by 'proof' testing developed in successive centuries, and today we have 'proof and experimental establishments' like that at Shoeburyness as historic links with earlier days. The structural strength and safety of shotgun barrels are still tested today by proof testing, where a charge well in excess of that normally associated with the explosion of a cartridge is fired in the chamber of the gun.

As the design of guns progressed, early versions of 'thick cylinder' theory began to be used to predict strength, and this was probably the first use of structural analysis to judge response to shock loading. The earliest guns fired simple stone or cast-iron cannonballs, or canisters containing grape shot or flint pebbles. The destructive power of these weapons was mainly due to collision impact between the balls and the outer fabric of structures, whether stone structures such as castles or walls, or the structure of the human body. However, a major and far-reaching development occurred in the fifteenth and sixteenth centuries when the explosive shell was invented by the French. The solid cannonballs were replaced by hollow spheres of iron containing a fuse, a bursting charge, and smaller fragments. This meant that explosions now took place when the projectile made contact with, or penetrated, the target. For the first time it became necessary to predict the response of heavy stone or rock fortifications to contact explosions. Later the projectiles were elongated and formed into the now recognizable shapes of bombs, mortars, armour penetrating and air-bursting shells and rockets. Much of the inventive British development in this field occurred at the beginning of the nineteenth century, due to illustrious men like Sir Henry Shrapnel and Sir William Congreve. The explosive shell or bomb became the major attacking weapon against the structures of land fortifications, naval vessels and military targets. The designers of fortifications had to analyse much more closely the response of masonry to localized explosions. Repeated hits on small areas of masonry could lead to breaching, and explosion after penetration could cause fragmentation, cracking and collapse.

Although there was much development in the science of demolition and in the effect of shock waves on the structure of stone and masonry, the analysis of structural response as we now know it began with the invention of reinforced concrete and the rapid growth of steel- and concrete-framed building structures in the early years of the twentieth century. This growth coincided with two of the most dreaded inventions of man, the military aircraft and the aerial bomb. Thus in the First World War the main structural damage on land was of traditional building structures due to the high-explosive artillery shell, but in the Second World War the structural problem had shifted to the effect of bombs on framed structures. In Britain many of the outstanding engineers and scientists of the day came into the field in support of the Ministry of Home Security. Professor John Fleetwood Baker (later Lord Baker), Professor William Norman Thomas and Professor Dermot Christopherson (later Sir Dermot Christopherson) wrote papers and reports of great significance at the end of the Second World War on structural response. It was realized that design methods involving elastic stress distribution for static structures were not suitable for analysing the effects of blast loading, and that the critical factor was the capacity of a structure to absorb energy. The ability to absorb energy in the plastic range was seen to be very significant in comparison with the capacity in the elastic range for most types of ductile structure. The carrying capacity of simple beams made of steel when bending beyond the limit of the elastic distribution of stress had been examined earlier by Ewing in Britain and Maier-Leibnitz in continental Europe. Their work was extended and the so-called 'plastic method' was used in the design of basements, surface shelters, indoor shelters and factory walls to withstand the effects of blast energy from closely exploding bombs, and to give protection against bomb fragments, crater debris, and other objects projected violently by the explosion.

Much full-scale test work was undertaken during the wars of the twentieth century by most of the major participating nations. The need to calculate the charges required to demolish structures, particularly reinforced and unreinforced concrete structures, became important, and most countries produced simple formulae linking the dimensions of concrete protective structures with the weight of the explosive charge and the internal volume of air space. In steel and wrought-iron building frames the possibility of progressive collapse was investigated, and the need for alternative load paths became an important consideration. Problems due to weakness in the connections of framed structures were evident, and designers were reminded that the more nearly that connections could be made to approach the continuity and ductility of the main members they join, the less the damage that would result from a given bomb explosion. The framework needed to be capable of resisting collapse if one main member was suddenly removed, and in no circumstances should it be liable to progressive failure.

The modelling of structures as simple mass-spring systems and the comparison of the natural periods of oscillation of such systems with the duration of the pressure pulse of an explosion were important in the determination of the level of structural response before the main energy absorption activity began. The advent of computer-based software in the mid-twentieth century has enabled quite sophisticated modelling to be undertaken in an effort fully to understand the physics of blast-structure interaction. More complex, multi-degree of freedom systems can be investigated quickly by the designer, and the influence on behaviour of structural damping assessed. This has led to a great increase in the computer-based analysis of structures subjected to shock loading. It has also become much easier to investigate theoretically the diffusion of shock waves in structural materials, and to use this to predict material behaviour resulting from explosive shock characteristics. The spalling and scattering of steel and concrete, and localized fragmentation at the point of application of the impulsive loading, can be investigated.

It was clear from the earliest time that structural damage would be considerably increased if the explosion of a charge took place within the material of the structure, so the penetration of attacking missiles into structures followed by subsequent explosion became an important aspect of design. It is not possible to discuss structural loading fully without discussing penetration, whether it is penetration of concrete by free-falling bombs or of the armour plating of military vehicles, naval vessels or military aircraft by armour-piercing shells, bombs, rockets or missiles. The penetration of high-velocity objects into soils, stone, metals and concrete has historically been a subject for military engineers, and much of the work in this field originated in the military establishments of Europe in earlier centuries.

The most famous of the early contributors to this field was Jean Victor Poncelet, an illustrious product of the French Ecole Polytechnique system, who was born at Metz in 1788. Part of his career was spent as a military engineer in Napoleon's Army, and part as an expert in engineering mechanics in the arsenal in Metz. He made outstanding contributions to the fields of geometry, structural dynamics, and engineering statics. He is said to have introduced the effect of shearing force into the calculation of beam deflection. His influence on the subsequent development of structural analysis was profound, and among his most far-reaching research was the study of penetration. He saw that the kinetic energy of penetrators, proportional to the square of the velocity, was a key ingredient of the equation, and that the flow of soils or metals around the body of the penetrator could not be ignored.

There are two aspects of penetration mechanics relevant to our subject. The first of these is the penetration of an explosive-carrying missile into the target medium before the explosion occurs. The second is the penetration of structures by high-velocity fragments of the metallic casings of bombs and shells after the explosion has taken place at a short distance from the target. The first aspect results in internal shock waves in the material of the target, the second results in high-velocity impact damage.

High-velocity impact from the steel fragments of shell casings began to be investigated after the advent of the artillery shell, and much work was in progress in the nineteenth century. However, the penetration of warheads and explosive-carrying devices did not become widely researched until the beginnings of aerial bombing in the twentieth century. The influence of penetrator shape on the depth of penetration, and the relevance of the strength of the penetrated material was examined, and at the end of the Second World War there were several empirical penetration formulae available. Most of these assumed that the penetrator did not deform on contact with the medium. The problem of a deforming plastic penetrator seems to have been considered firstly by G. I. Taylor, but in recent times there has been an increase in research in this area resulting from the development of three-dimensional finite element programs capable of dealing with dynamic forces and massive plastic deformation.

Once the fundamentals of deformation, blast loading, structural response and penetration have been examined the problems facing the structural designer in many fields can be addressed. The first of these is the effect of local explosions, where up to one or two tonnes of explosive is detonated on or near structures. The targets might be aircraft structures, naval structures, underwater structures, protective structures built on the surface of the ground, or protective structures built below ground. We owe much to the research and testing aimed at the evaluation of structural response to 'conventional' or 'non-nuclear' bombs carried out in the USA, Great Britain and Germany during the past 30 years. The leading American establishment in this work has been the US Army Waterways Experiment Station at Vicksburg, Mississippi. Their test facilities have been constructed at great expense, and have been used to examine explosions/ structure interaction over a vast range of conditions. In an earlier book, *Buried Structures*, the author has reviewed their test rigs and the results of their research on the effect of local explosions on underground thick-walled and thin-walled structures.

Much of the experimental work has been aimed at reinforced concrete structures, because so many of the protective structures that are built to withstand the effect of conventional bombs or shells are constructed from this material. In earlier times the structures of castles and forts were exclusively of stone, or of combinations of stone and earth. The Romans, for example, in their city fortifications used two strong walls of masonry separated by a gap of six metres, and the gap was filled with well rammed earth and loose rock. By the fifteenth century the advent of the cannon firing balls over 30 lb in weight meant that a wall of masonry two or three metres in thickness could not resist the effects of a battery discharging three to five hundred balls over a surface about eight metres square. This led to the development of 'relieving' arches within the body of the wall and internal abutments buried in the adjacent earthworks. Often the external surfaces of the walls were further protected by ramparts of earth and wood. This increase in strength proved effective until the coming of the exploding balls and shells, which was discussed earlier. The local explosion in the midst of the ramparts caused havoc and disorder, and this led to the replacement of castle and wall by the bastioned fortress. During the seventeenth century there was an absorbing conflict between the developers of such protective structures as the 'star' fortress and the creators of systematic methods of attack. The genius of the age was Sebastien le Prestre de Vauban, Marshal of France, who became the outstanding technical director of siegecraft and fortress defence at the same time. He also built bridges, roads and canals, and created the French Corps of Engineers in 1676. In the eighteenth and nineteenth centuries warfare was brought more into the open, and commanders such as Napoleon and Frederick the Great avoided attacks against permanent defences. However, the invention by the French at the end of the nineteenth century of smokeless powder and high-explosive artillery shells greatly increased the attacking power and the need for strong land fortifications. This coincided with the rapid growth of concrete as a replacement for stone, earth and masonry.

Later, the use of thick reinforced concrete construction was vital in resisting the effects of direct hits from aerial bombs weighing 1000 kg and above. The French carried out field tests during the construction phase of the Maginot Line before the Second World War, and they deduced that 3.5 m of reinforced concrete could withstand three direct hits at the same point of impact. The ultimate fortifications of this period were the massive U-boat pens built by the German Army in Brittany to protect their submarines against aerial bombs. After the end of hostilities in 1945 the USA conducted a Strategic Bombing Survey, and the reports of this work contain a record of interviews, reports and photographs, including the effects of the conventional bombing of German cities.

Very detailed reports of bomb damage to reinforced concrete and steel building frameworks were published in Britain, and from these and other sources enough information was gathered to produce for the first time design guidance for engineers who needed to calculate the response and strength of structures that were specified to be resistant to accidental or man-made local explosions.

Another military threat about which much has been written is the nuclear explosion. Although the threat seems less immediate these days, the major arsenals of the world still contain nuclear bombs, and the behaviour of structures in the shock and blast wind phases of a nuclear explosion must still be assessed by engineers. It is the duration of the dynamic pressure, or drag loading, that is the main difference between nuclear and high-explosive detonations. The duration of the positive phase of the dynamic pressure from a megaton nuclear explosion can be several seconds, whereas the duration of the air blast from a conventional high-explosive detonation may only be a few milliseconds. Structures most likely to be damaged by the high instantaneous pressure associated with shock front are dwelling houses. Structures likely to be damaged by the dragforce of the blast winds are chimneys, poles, towers, truss bridges and steel-framed buildings with light wall cladding. There are also the hazards of fast-flying debris and fire. The threat is so great that nuclear resistant structures are normally buried below the ground surface.

A great deal of information about the behaviour of structures of all types was assembled after the Second World War from the controlled nuclear bomb tests in the Pacific and at the US test site in Nevada during the 1950s. Publications by Professor Nathan Newmark and others on the design of structures to withstand nuclear effects, including the problem of radiation, were important and progressive contributions to the structural mechanics of the problem. The American Society of Civil Engineers was particularly active in this work. The importance of underground structures led to a surge during the 1960s in analytical and experimental research on soil/structure interaction in a dynamic environment, and a number of simulation facilities were built in America and Britain. As the political problems of detonation and fall-out from field nuclear tests increased, nuclear bomb effects had to be approximately simulated by exploding a great weight of TNT instead. Unfortunately much of the target response information is hedged in by a high-security classification, and the details are not freely available.

The effect of the heat flash associated with a nuclear explosion can also damage certain types of structure, particularly when the structure is made of aluminium. It is possible for aluminium military equipment, such as a rapidly built bridge, to escape damage by pressure or wind, but be subjected to temperatures that are high enough to reduce the strength of the alloy. The other hazard, as mentioned above, is due to initial nuclear radiation. As a rule structures designed to protect occupants against peak overpressures of 77 KPa and above should also be checked for radiation. Radiation can be particularly damaging when peak overpressures exceed 315 KPa. The response of structures and the humans in them to nuclear explosions is therefore a very complex design field, and the engineering of the subject is often left to physicists, with questionable results.

Another field of interest is what might be termed civil explosions. These are non-military or non-terrorist explosions which are either the result of accidents due to natural forces or deliberately produced to aid the requirements of civilized society. The most frequent accidental explosions are probably due to domestic gas leaks, and there have also been accidents due to the ignition of methane gas in civil engineering works and mines. In Britain much structural research has centred on the former Building Research Establishment. Many reviews and reports have been written on the nature of serious gas explosions and structural disasters are well documented. The action of domestic gas explosions is nearly always from within the structure, and in structures that are badly designed with respect to progressive collapse a relatively small explosion can lead to spectacular results. The collapse of a high stack of corner rooms at the Ronan Point high-rise apartment building in London following a gas explosion led, in the 1960s, to a formal enquiry and adjustments to future design requirements for all types of building. Internal gas explosions are unfortunately still a regular occurrence, often resulting in the death of house or apartment dwellers.

The possibility of internal explosions also occurs in buildings that house test facilities for engines, and here it is good practice to limit the instantaneous pressure rise of the explosion by using weak panels in the superstructure that will collapse outwards. Other industrial manufacturers who require protective construction in their facilities are the producers and storers of explosive materials. The possibility of an accidental explosion causing detonation of other explosive devices is a hazard that has to be carefully examined at the building design stage. Hardened buildings that limit the propagation of mass detonations are usually constructed of reinforced concrete with steel blast doors. There are regular conferences, particularly in the USA, on protection against the accidental explosion of hazardous mixtures, and safety control in this field is becoming increasingly important in most countries.

Another civil hazard is the unconfined vapour cloud or dust cloud explosion, which has often resulted in a major national disaster. In Britain the on-shore explosion of the Flixborough chemical plant and the off-shore explosion at the Piper Alpha oil rig have caused much damage and loss of life. Explosions of this type cause secondary damage due to fragmentation, fire and loss of structural stability. There is always much publicity and speculation when large cloud explosions occur, and the political pressures to find the causes and distribute the blame are usually as intense as the explosion itself. Smaller explosions of this type have occurred from time to time in ships, under cargo conditions that are not always foreseen as dangerous. The problem of design against accidental explosions has been addressed by the US Department of Energy, who commissioned the writing of a manual for the prediction of structural loading.

We do not usually think of quarry blasting as a structural response problem, but of course the fissuring and breaking up of the structure of stone and rock is just that. What is more important is the increasing use of explosives to demolish civil structures. In the past this subject has not been considered worthy of detailed scientific research, and most of the effective and safe demolition methods have been properly learned by experience. Publications in this field have tended to be of a handbook nature, and none the less valuable for that, produced by the large explosive marketing corporations like Dupont or ICI. There is, of course, much military knowledge on the subject, particularly in the effective deployment of shaped charges to cut through the metal members of bridges and building frames. The military engineers are also interested in the way a bridge collapses; they prefer to twist as well as break the structure so that the collapsed bridge cannot easily be used as a basis for a temporary crossing.

Controlled explosions can be used to aid the decommissioning and demolishing of nuclear reactors and off-shore oil platforms. There is also the possibility of using explosions to demolish prestressed concrete structures, where the danger lies in the energy stored in the structure by high forces in the tendons. Another civil area is the effective design of bank safes against the use of local explosions by professional burglars; this is not considered a terrorist activity as human life is normally spared, but for insurance purposes it is necessary to introduce scientific methods into safe design. It is not easy to obtain much information about this expertise because publicity would provide too sound a basis of technical knowledge for the safe-breaker, and there have only been one or two conference papers on the subject.

Finally, we must consider the assessment of the residual strength of damaged structures. This is an important subject in the civilian world where public safety has to be guarded, and in the military and industrial world where the effect of damage on future operations needs to be examined. There are two particularly important areas. The damage by terrorist bombs to the structure of buildings often means that immediate temporary measures must be taken to shore up walls, support lintels, or repair joints.

The second area is the damage to the structure of aircraft and ships by local explosions and fragmentation impact. Military aircraft and some naval craft are particularly vulnerable to battle damage, and methods of assessing the residual strength of plates, members and sandwich construction due to penetrating weapons are of operational interest. The damage may be of a denting or crumpling nature, caused by blast winds or fragmentation. A good deal of attention has been paid to this area of the subject by the Advisory Group for Aerospace Research and Development of the North Atlantic Treaty Organisation as part of their investigation of the impact damage tolerance of structures. Historically aircraft of war returned home on a wing and a prayer, but the feeling in recent decades is that the prayer should be replaced by a microchip that monitors the damage as it occurs and gives instructions to the controlling computers.

It can be seen from these introductory remarks that there is a great deal of data available and, as we said earlier, conferences on the subject are taking place with increasing regularity. The aim is to review the history as concisely as possible, beginning with the physics and ending with the engineering of the subject. The future is not easily predicted. Nuclear explosions are less likely, one hopes, but local explosions due to terrorism or revolution seem to be growing in number throughout the world. The need for research and development will no doubt remain, perhaps with a change of emphasis from military to civil requirements. There is already a move to increase the amount of university funding in this area, underwritten by the response of senior committees to the need for fuller research in support of the 'new' subject of Hazard Engineering, and this text may help to introduce the problems of explosions and structures to a wider audience of academics, consultants and advanced students.

1

The nature of explosions

1.1 THE DEVELOPMENT OF LOW AND HIGH EXPLOSIVES

Historians of science tell us that the explosives industry began with gunpowder, made by hand-mixing potassium nitrate (saltpetre), carbon (charcoal) and sulphur in a chemist's mortar. Most schoolboys are familiar with this basic mixture from their chemistry lessons, where they were taught that the ideal proportions should be based on the atomic weights of the components, about 75 : 13 : 12. However, the gunpowder of the thirteenth century, developed by the English philosopher and scientist Roger Bacon, was rather short on the saltpetre, and his proportions were about 40 : 30 : 30. It is of interest that Bacon was born near Ilchester in Somerset and that he turned to science and philosophy when forced to retire from his work as an academic. There is hope here for all disillusioned academics! Like so many creative inventors his work was marked by a foresight based on imaginative speculation, in which he combined the cross-reading of various sciences with a knowledge of the ways of history and human error.

The shortfall on saltpetre was probably because this compound was not available in a sufficiently refined form at the time of Bacon, but once it had been shown that after ignition his mixture would burn with great speed, and give rise to large quantities of gas, the military soon became interested. Following ignition in a confined space large pressures could be used to propel missiles such as lead shot and cannonballs. Black powder, named because of the charcoal content, was quite dangerous to handle because of the possibility of spontaneous ignition, but this did not stop the mechanization of the mixing process over the centuries. Batteries of pestles and mortars were powered by horse or water power, but by the end of the eighteenth century in England there had been so many fatal accidents that these 'stamp-mills' were forbidden. It was also found that ramming the powder into the breaches of guns was a very tricky business. If packed too tightly burning was so slow that there was an insufficiently rapid build-up of propelling pressure; if packed too loosely most of the gases escaped around the projectile and the subsequent speed of travel of the shot was much reduced. This problem was overcome by mixing the powder with fluids to form a cake, and then breaking the cake into uniformly sized crumbs or grains through a sieve. Since two of the best fluids were alcohol and urine, the making of gunpowder was not unlike a farming process, and as the original scientist in the field had links with bacon and Somerset cider this was not surprising. Incidentally there were still members of the Bacon family living in Yeovil, the author's hometown, about five miles from Ilchester, in the mid-nineteenth century, and the late-twentieth-century telephone directory shows a number of Bacon entries in the Yeovil, Crewkerne, Taunton and Bridgwater areas of Somerset.

The search for saltpetre was a serious activity over the centuries. In the early days it was known that there were surface deposits in Spain and India, but in most European countries no ready-made supply existed. Consequently the business of 'nitre' beds prospered, in which layers of decaying animal and vegetable matter, earth, sand and old mortar were moistened from time to time with blood. Eventually potassium nitrate could be extracted. This method was mainly overtaken when vast deposits of sodium nitrate, which could be converted to saltpetre, were found in Chile. The military use of black powder, with its powder kegs, powder horns and the commands to 'keep your powder dry' was much reduced at the end of the nineteenth century, when 'smokeless' powder was produced in France. Apparently one of the last non-military uses on a large scale was in the destruction of a huge rock in New York harbour (the Pot Rock) in 1853. We are told that 200 000 lb of powder were used. Gunpowder was also used in the mining of ore in Europe from the seventeenth century, and for road widening in Switzerland at that time; and of course we all know of the 36 barrels, weighing about one cwt each, concealed beneath coal and faggots under the House of Lords in 1605.

The firing of black powder produced so much smoke that after a heavy volley from land-based guns or a broadside from naval guns the firers and the target became completely obscured. A new propellant, smokeless and without residue, was needed, and this led to the development by the French scientist, Paul Vieille, of a mixture of black powder and gelatin which was smokeless. The action of all powders upon initiation is by burning, or deflagration. Each grain of the powder burns at the same time as other grains and the internal pressure throughout the mass of powder remains uniform and equal to the external pressure. It is possible for the deflagration to change to a detonation in which the chemical reaction spreads like a wave from the point of initiation, a change that is completed in a few millionths of a second; but smokeless powders are very difficult to bring to this stage. They are therefore 'low' explosives.

Most other conventional explosive substances have been developed from nitric acid and nitrates, and the distillation of potassium nitrate, alum and blue vitriol to form nitric acid is usually considered to have originated with the Arabian chemist Greber (or Jabir) in the eighth century. He was the most celebrated chemist of medieval times, a skilled experimentalist and a developer of theories with their roots in Greek science and occult philosophy. Almost a thousand years later the production of nitric acid had changed to a process invented by the German chemist Johann Rudolf Glauber, who was born at Karlstadt in Bavaria. He eventually settled in Holland and built a magnificent laboratory in Amsterdam in the mid-seventeenth century where among other things he defined the medicinal properties of Glauber's salt. His process for nitric acid was to heat a nitrate with concentrated sulphuric acid, but nowadays there are several commercial processes, the most common of which use the catalytic oxidation of ammonia. Nitric acid is very powerful and has formed the basis of most modern explosives. In 1838 T. J. Pelouze discovered that cotton could be converted into an extremely inflammable substance by the action of concentrated nitric acid, and in 1845 it was shown by C. F. Schönbein that this substance could be made into an explosive by adding sulphuric acid to the nitric acid. Nitrocellulose was introduced as an ingredient of gunpowder in the 1860s, and when E. A. Brown discovered in 1868 that nitrocellulose could be exploded by a detonator the substance began to be used as a high explosive. It is an unstable material, however, and stabilizers need to be added to convert the substance into the more reliable explosive known as Guncotton. This is a white, inodorous, tasteless solid that retains the structure of the cotton.

A much more far-reaching development based on nitric acid was the discovery in 1846 by the Italian scientist A. Sobrero of nitroglycerine, made by the treatment of glycerine with a mixture of concentrated nitric and sulphuric acids. It was the study of nitroglycerine that introduced the name of Nobel, first to the explosives industry, then to oil production, and then to the world via the Nobel foundation and its prizes. Alfred Bernhard Nobel, engineer and chemist, was born in Stockholm in 1833, and by 1867 he had combined nitroglycerine with a diatomaceous earth, known as kieselguhr (kiesel is gravel, hence Chesil beach in Dorset), to form dynamite. This made the commercial use of nitroglycerine possible and was an invention of great importance. The patenting of Guhr dynamite, as it was called, in Britain and the USA was followed by a number of other dynamites in which nitroglycerine was mixed with other substances capable of sustaining an explosion. These included sodium nitrate, wood pulp, calcium carbonate and charcoal. In 1876 Nobel patented blasting gelatin, which was the most powerful and shattering explosive of its time, made by dissolving 7% of collodion nitrocellulose in nitroglycerine. In 1889, a few years before his death in 1896, Nobel produced ballistite, a nitroglycerine smokeless powder which was the forerunner of cordite.

Of equal importance to his work on dynamite, Nobel also constructed the fulminate of mercury detonator for initiating the explosion of nitroglycerine and guncotton. Fulminate of mercury became known as an initiating or primary explosive, because it could easily be detonated by impact, and could therefore change ignition into detonation by shock rather than heat. These developments, together with exploitation of the Baku oilfields, brought Alfred Nobel a fortune,

or rather an immense fortune. He was unmarried, dogged by ill-health, pessimistic and cynical, but luckily he combined these features with a powerful
benevolence. When he died in 1896 he left most of his fortune in trust, to form
the Nobel foundation from which international prizes are awarded annually for
Physics, Chemistry, Medicine, Literature and Peace.

The end of the nineteenth century also saw the discovery of a range of
military explosives of great importance. It was found that methyl benzene
(toluene) reacted with nitric acid in the presence of concentrated sulphuric acid
to form trinitrotoluene, more commonly known as TNT. This became the
standard explosive of the First World War, and since then the combination of
liquid TNT with explosives such as RDX or PETN in plastic form has proved
very successful. TNT can be manufactured with relative safety and economy,
and because of its universal use it has become customary to class all types of
explosive (conventional or nuclear) in terms of TNT as a standard.

PETN is pentaery-thritol-tetranitrate and RDX is cyclo-trimethylene-trino-
tramine. HBX is a combination of RDX, TNT and Aluminium; pentolite is a
mixture of PETN and TNT in equal parts. Other important families of high
explosive were the Amotal and Ammonal groups, formed by compounding
ammonium nitrate with TNT. Many combinations of nitrates, TNT and aluminium were tried, giving rise to a range of trade names that need not concern us.
Explosives of all the above types were the main materials for bombs in the
Second World War, and extensive wartime research established a league table of
strengths in terms of peak instantaneous pressure and positive impulse.

The main development in the latter half of the twentieth century has been
'slurry' explosive. This began with the mixing of ammonium nitrate and fuel oil
to form ANFO which, although extensively used for quarrying and mining, was
somewhat low in strength and susceptible to the effects of water. These
problems were overcome by combining the ammonium nitrate with sodium
nitrate in a gelled aqueous solution to form a slurry. This is widely used because
it reduces the cost of many of the civil needs for high explosive, and there has
been a noticeable reduction in the use of the nitroglycerines in recent years. For
military purposes, however, the production of TNT-based explosives in sheet
and slab form has continued, and most readers will be aware of the use by
terrorists of sheets of the Czechoslovakian explosive, Semtex.

Military research establishments have also been involved in the development
of fuel-air explosive (FAE), a detonating explosive that draws virtually all of the
required oxygen from the air. Detonation occurs over a large area so that a
greater impulse is generated than with the point detonation of a conventional
high explosive. Research shows that about 42% of the weight of TNT is due to
the oxygen attached to the explosive molecule $12C_6H_2CH_3(NO_2)_3$, whereas 47%
of the weight of the consumables (fuel and air) in some types of fuel-air
explosive comes from the surrounding air, and does not have to be carried
within the body of the explosive. This means that a given weight of FAE
produces over seven times the energy of TNT, although the peak overpressure is

much lower and the duration of the impulse much greater. Detonation, as opposed to deflagration, is obtained by the use of a burster charge containing a mass of conventional high explosive about one-hundredth of the FAE mass. The control of FAE weapons was thought to be uneven, and research in the USA was reduced in the 1970s, but more recently the Soviet air force used weapons of this type in Afghanistan, and there are reports in the technical military press of the very successful use of FAE bombs by the US forces in the Gulf War.

Let us now turn to the detonation process, which has been considerably developed during the past century. There are now many methods for the safe detonation of high explosives, and nearly all consist of the initiation of a shock wave into a base charge which then detonates the main explosive charge. Most recruits to the armed services in the first half of this century will have handled the small aluminium tubes containing a priming charge of mercury fulminate ($HgC_2N_2O_2$), or lead azide (PbN_6) and a secondary charge of PETN. The primary charge could be set off by impact, as in the spring released system of a hand grenade, or via a length of safety fuse capable of initiation by a safety match.

The priming charge can also be ignited by electrical means, and electrical shotfiring is now widely used. When an electrical charge is connected to the detonator it heats up a fine wire which then ignites a flashing compound. This 'fusehead' fires the primer and base charges. Delaying elements can be introduced between the fusehead and the primary charge, or the fusehead can be set to explode so quickly that the entry of the firing current and the detonation of the base charge is virtually instantaneous. The use of electrical detonation must be carefully controlled in very sensitive environments where there is a risk of premature firing by static electricity, or where there are dangerous methane or dust clouds. Special detonators have been developed to overcome these problems and to avoid accidental pre-detonation. These are described in handbooks and in the marketing literature of manufacturers such as ICI Nobel and Dupont.

The other main detonating medium is the 'detonating cord' and 'Cordtex' is a universally used version in blasting and quarrying operations, explosive cutting, and underwater operations. A core of PETN is surrounded by tape and wrapped with textile or synthetic yarn, and this cord is completely enclosed by a white plastic tube. It has a high velocity of detonation which means that it will initiate most commercial explosives, and it can be initiated itself by most of the standard detonators. There is plenty of information on methods of using Cordtex for various blasting operations, involving detonating relays, jointing and branching methods, and initiating systems.

A major area of development has been the artillery fuse. These were originally the powder train fuses, or rather inaccurate mechanical devices, but more recently the need for high firing rates has led to the evolution of the electronic time fuse and to the electronic fuse with a high-explosive boosting charge. The fuse is the shaped end of a mortar or artillery projectile and consists of an ogive, an electronic assembly, a power-supply assembly, a liquid crystal display assembly, and a safety and arming device. Once the weapon is fired, the electronic assembly senses continuous spin above a designated speed in revolutions per minute, and then initiates the fusing operations. These consist of the activation of the safety and arming device which makes the fuse ready to function upon target impact. When the fuse functions the explosive in the projectile detonates. Electronic technology has now advanced to allow the fitting of a proximity fuse within the same standard configuration, so that an electromagnetic signal is generated which reflects off the target area to initiate detonation at a specific height above the target.

There are numerous technical exploration programmes to improve the electronic fusing of all types of missile warheads. The problems of early detonation or battery failure, the consistency of burst height, rain sensitivity, and vulnerability to electromagnetic effects are all areas where the research level is high. The type of target is an important factor. Urban targets such as triple brickwork and concrete bunkers need different delay, proximity and time characteristics in fuses than shipboard armour penetration. The search continues for effective multi-option fuses that can be normally or inductively set and used for a range of targets and a range of weapons, projectiles and propelling systems.

1.2 NUCLEAR EXPLOSIONS

At about the time the development of TNT as a high explosive began, the French physicist H. A. Becquerel noticed in 1896 that uranium emitted unusual radiations, an observation of great significance that came as a consequence of the discovery a few months earlier of X-rays by W. K. Röntgen, professor of physics in the University of Wurzburg, Bavaria. Becquerel's original investigations on the photographic effects of penetrating radiation were quickly taken up, the science of radioactivity was established, and by 1899 the New Zealand-born scientist Ernest Rutherford had shown that the radiation from uranium was a complex matter, involving easily absorbed radiation (α rays), and more penetrating radiation (β rays). Later the extremely penetrating γ rays were discovered.

The particles were used by Rutherford to explore atoms over a period of many years, and in 1919 he observed that nitrogen, when bombarded by particles emitted the nucleus of the hydrogen atom. This nucleus became known as a proton, and with its emission the atom had been smashed. More nuclear transformations were discovered by J. Chadwick in 1932, when he recorded the neutron, capable of freely penetrating many centimetres of material. His work was taken forward, and the utilization of nuclear energy on a large scale was made possible just before the Second World War when the scientists Hahn and Strassmann from Berlin discovered uranium fission. The impinging of neutrons on the uranium nucleus produced a number of radioactive substances because the nucleus was on the verge of disintegration before the bombardment began. The breaking up of the nucleus was named the fission, and in 1939 Szilard,

Joliot and other experimentalists succeeded in liberating enough neutrons in the fission process for a chain reaction to develop. Under controlled conditions the fission became an energy liberator, as demonstrated in 1942 by Fermi of Chicago and under uncontrolled conditions the chain reaction became an explosive of unimagined force. Using this principle, a team led by J. R. Oppenheimer exploded the first 'atomic' bomb near Alamagordo, New Mexico, in July 1945. The effects on the conclusion of the Second World War were far-reaching, two atomic bombs were dropped on Japan and in a month the conflict was over. Further research followed. It was found that fission, the splitting of large atoms of heavy metals, could be replaced by fusion, the fusing together of atoms. The fusion reaction was made easier if the nuclei involved were hydrogen nuclei, and if the temperature of the reaction was extremely high. These factors were the basis for the production of 'hydrogen' and 'thermonuclear' bombs.

The development of nuclear energy and nuclear weapons illustrates the gathering speed of scientific research and development in the twentieth century, due mainly to the rapidly improving means of human and written communication. The first atomic bomb was exploded only fifty years after the original discovery of radioactive uranium by Becquerel. It is also worth noting that although the scientific discoveries took place in the major countries of Europe, the translation of this research into practical forms needed a country with the resources and wealth of the USA.

The bombs dropped on Hiroshima and Nagasaki were each equivalent in power to about 20 000 tons of TNT, but there soon followed experiments with thermonuclear bombs many times greater in strength. Nuclear bombs are often detonated by high explosives which are themselves detonated by initiating explosives. The radioactive fallout that accompanies the blast and shock of the explosion is an added hazard that makes the nuclear bomb a particularly hated weapon. No nuclear wars have taken place since 1945, but we cannot ignore the effects of such an explosion in our survey of structural loading because, as stated earlier, so many countries have nuclear weapons in their arsenals.

The explosive actions of high explosive and nuclear fission are similar in that the release of energy happens so quickly that the sudden expansion compresses the surrounding air into a dense layer of gas. This dense layer expands so fast that it forms a shock wave, the face of which is known as the shock front. It moves supersonically, in contrast to a sound wave that moves at 'sonic velocity' and does not 'shock-up', to use the jargon of the explosive scientist. The instantaneous rise of pressure at the shock front falls away and declines gradually to sub-atmospheric pressure. It is the physical analysis of this action that we shall be reviewing.

The relationship between the surface area and the mass of a fissionable isotope of uranium is an important characteristic in the detonation of a nuclear explosion. If the ratio of area to mass is large too many neutrons will escape for the fission chain to be produced, and the mass is said to be 'sub-critical'. Clearly for an explosion to take place the weapon must contain a 'supercritical' mass, and this must be produced in a very short space of time, either by firing one sub-critical mass at another using explosive propellant, or by compressing a sub-critical mass by the implosion forces of surrounding high explosive. Information on the design of nuclear weapons was a matter of high security for many years, but recently more information has become available on the size, shape and action of the devices.

1.3 THE ANALYSIS OF DETONATION AND SHOCK IN FREE AIR

Before discussing the effects of explosions in the various constituents of the physical world, such as air, water and earth, we must survey the build-up of theoretical knowledge by scientists since the nineteenth century. Once high explosives had been developed it was natural that attempts would be made to predict the detonation behaviour and to deduce mathematical expressions for the energy of explosions and for the propagation and decay of blast waves in air.

It was noticed that detonation propagates as a wave through gas in a very similar way to the propagation of a shock wave through air, so the work of scientists on the physics of shock waves became important. It was inspired to a great extent by earlier work on the theory of sound and sound waves, by Earnshaw [1.1] in 1858 and Lamb [1.2], the English mathematician (1849–1934), who became the acknowledged authority on hydrodynamics, wave propagation, the elastic deformation of plates, and later on the theory of sound. The publication of his works on the motion of fluids coincided with the growing interest in explosions, and his contribution was discussed in the Introduction.

The conditions relating to pressure, velocity and density in a gas before and after the passage of a shock wave were first investigated by Rankine [1.3] in 1870, then by Hugoniot [1.4] in 1887. Their work was used a few years later by D. L. Chapman, who suggested that a shock wave travelling through a high explosive brings in its wake chemical reactions that supply enough energy to support the propagation of the wave forward. At the same time the French scientist J. C. E. Jouget suggested that the minimum velocity of a detonation wave was equal to the velocity of a sound wave in the detonation products of the explosive, which were at high temperature and pressure. All this work, described recently by Davis [1.5], applied strictly to explosive gases, but was assumed to apply to liquid and solid explosives. There was further research on the subject in Russia, the USA and Germany in the 1940s, but the most useful analysis of the detonation process was set down by G. I. Taylor [1.6] in a paper written for the UK Civil Defence Research Committee, Ministry of Home Security, in 1941, during the Second World War. He took a cylindrical bomb, in which the charge was detonated from one end, and in which the reaction might advance along the length of the bomb at a speed of over 600 m/sec if the charge were TNT. The internal pressure forces the casing to expand, the expansion being greatest at the initiating end. When the case ruptures the explosive gases escape and form an incandescent zone that expands so rapidly that a shock wave or pressure pulse is formed.

Taylor wrote many valuable papers [1.7], [1.8] on the dynamics of shock waves for the Civil Defence Research Committee in the early days of the Second World War, and it is from these that much of the analysis has been taken. His work was summarized lucidly by D. G. Christopherson in 1945 [1.9] and the latter's summary has been consulted frequently by the author in writing the chapters of this book. In the summer of 1945 Christopherson wrote his seminal report on the structural effects of air attack, the information having been collected during the Second World War by the Research and Experiments Department, Ministry of Home Security, and much of it coming from experiments carried out on behalf of that Department by the Building Research Station and Road Research Laboratory. Contributions were also drawn from the work of the National Defence Research Committee in the USA. Christopherson's report, entitled *Structural Defence*, covered every aspect of the subject from the theory of blast waves in air, earth or water to the general theory of structural behaviour and the design of protective structures of all types. Christopherson was educated at Oxford, was a fellow at Harvard and then a postgraduate at Oxford before joining the Research and Experiments Department of the Ministry of Home Security in 1941. He left, after writing *Structural Defence* in the space of two or three months after victory in Europe, to join the Engineering Department at Cambridge University in 1945. Since then he has pursued an illustrious academic career and has been a major influence in the teaching of Engineering subjects in the universities, as well as in the management and control of universities and colleges.

The form of the overpressure/duration relationship for a high explosive or nuclear explosion in air is shown in Figure 1.1, where p is overpressure (or air blast pressure) and t is time. In the figure the decay of pressure after the first instantaneous rise is expressed exponentially. There are other ways of denoting the form of the pressure/duration relationship, as we shall see later, but for the purposes of this chapter Figure 1.1 will be adequate. The value of the peak instantaneous overpressure p_0 will depend on the distance of the point of measurement from the centre of the explosion. Using imperial units, in a TNT explosion p_0 might be 200 or 300 psi at the point of burst of the explosion, but would rapidly diminish with distance. In a nuclear explosion equivalent to 1000 tons of TNT, the peak overpressure would be 2000 psi at 30 metres from the centre of the explosion. The duration of the positive phase is t_0 units of time.

Theoretically, for a perfectly spherical charge in air, the relationship between p_0, the distance of the point of measurement from the centre of the explosion (R), and the instantaneous energy release (E), takes the form

$$p_0 = KE/R^3, \tag{1.1}$$

so that an important non-dimensional parameter is $p_0 R^3/E$. In imperial units E is measured in ft lb, and in SI units in joules. Experiments show that the explosion

The nature of explosions

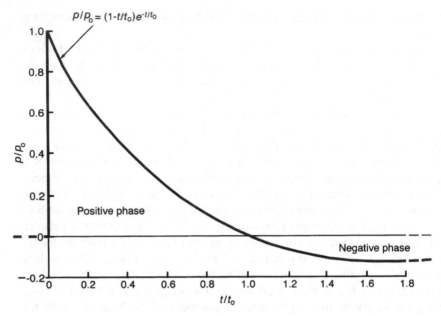

Figure 1.1 Overpressure/duration curve for detonation in air (idealization with exponential decay) (from Friedlander, ref. 1.19).

of TNT generates a blast energy of approximately 4600 joules/gram, which is equivalent to about 1100 calories/gram. In fact, the definition of a 'standard' gram of TNT is that which gives a blast energy of 4610 joules. The definition of a standard ton of TNT is an energy release of one million kilo-calories.

For a given type of chemical high explosive, energy is proportional to total weight, so it has become customary for design purposes to rewrite Eq. (1.1) as

$$p_0 = K_1 W/R^3. \tag{1.2}$$

K_1 is no longer non-dimensional and great care must be used in applying the formula. Experiments suggested that the equation did not give very accurate values of p_0 over the entire range, and an improved version was proposed in the US Army Technical Manual *Fundamentals of Protective Design (Non-nuclear)*, No. TM5-855-1, published originally in 1965 [1.10]. In imperial units the equation became

$$p_0 = \frac{4120}{z^3} - \frac{105}{z^2} + \frac{39.5}{z}, \tag{1.3}$$

where

$$p_0 = \text{peak pressure in psi}$$

$$z = R/W^{1/3} \qquad (R \text{ in feet, } W \text{ in lb}).$$

The relationship should only be applied when $160 > p_0 > 2$ psi, and $20 > R/W^{1/3} > 3$ ft/lb$^{1/3}$.

Note that the parameter $R/W^{1/3}$ enabled the results to be applied to any detonated explosion, conventional or nuclear, as long as the equivalent weight of charge in TNT were known. As we saw in the Introduction, $R/W^{1/3}$ is an important scaling factor, first noted by Hopkinson in 1915. Modern versions of Eq. (1.3) give the pressure in bars and z in metres/kilogrammes$^{1/3}$.

The relationship in Eq. (1.3) is very similar to the logarithmic plot of peak instantaneous pressure versus scaled distance given by Kennedy [1.11] at the end of the Second World War, when he summarized the results of free-air blast tests on cast TNT in the US and the UK. It was not realized at the time that the shape of the charge could have a significant influence on results, or that the pressure-measuring instruments could noticeably affect tests by altering the air flow behind the shock front. Nevertheless, Kennedy's summary was a useful indication of the pressure/distance relationship.

The pressure/distance characteristics discussed above only apply to a truly spherical charge in air, but in many practical circumstances the shape of the charge is cylindrical, or a plane sheet, or a line source such as detonating cord. For line charges it is known that the shock front expands cylindrically, so that p_0 is a function of $R(L/W)^{1/2}$, where L is the length of the source and $L \gg R$. Generally, blast waves from non-spherical sources exhibit a less rapid decay of pressure with distance. Theoretically predicted values of p_0 have been proposed by Lindberg and Firth [1.12], and are shown in Figure 1.2.

The vertical axis is p_0/p_a, where p_a is atmospheric pressure, and $E^{1/3}$ is replaced by the characteristic dimension R_0, where

$$R_0 = [E/(p_a L^{(3-v)})]^{1/v}. \tag{1.4}$$

$v = 1$, 2 or 3 for plane, cylindrical and spherical charges, respectively, so that for spheres $R_0 = (E/p_a)^{1/3}$, for cylinders $R_0 = [E/p_a L]^{1/2}$, and for line charges $R_0 = E/p_a L^2$). The overpressure ratio p_0/p_a is compared with R/R_0 in the figure.

Tests in 1948 on the behaviour of rectangular block charges of TNT, on TNT and Pentolite spheres, and on Pentolite cylinders were reported by Stoner and Bleakney [1.13], but there has been little published information on cylinders since then. Stoner and Bleakney gave the pressure-distance relationships in imperial units as:

$\frac{1}{2}$ lb rectangular blocks, TNT,

$$p_0 = \frac{36280}{z^3} - \frac{770}{z^2} + \frac{13.5}{z}, \tag{1.5}$$

8 lb cylinders, Pentolite,

$$p_0 = \frac{21070}{z^3} - \frac{135}{z^2} + \frac{10.5}{z}, \tag{1.6}$$

4 lb cylinders, TNT,

$$p_0 = \frac{19210}{z^3} - \frac{186}{z^2} + \frac{11.34}{z}, \qquad (1.7)$$

3.75 lb spheres, Pentolite,

$$p_0 = \frac{7823}{z^3} - \frac{295}{z^2} + \frac{8.63}{z}. \qquad (1.8)$$

For most of these formulae $p_0 < 2$ psi and $z > 20$ ft/lb$^{1/3}$. These different relationships were empirical and tended to suggest an accuracy of scientific measurement that did not exist. In fact, when the results are plotted on common axes, they all lie within a narrow scatterband, as shown in Figure 1.3. Further information on the behaviour of spherical charges of Pentolite is recorded in the works of Goodman [1.14], who compiled measurements taken from experiments over the 15 years from 1945 to 1960.

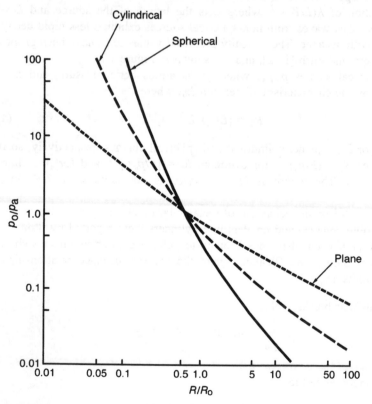

Figure 1.2 Peak overpressure v Range for various charge shapes (from Lindberg and Firth, ref. 1.12).

The results of much interesting research undertaken on behalf of the UK Ministry of Home Security during the Second World War are buried in the archives of the Public Records Office at Kew. From this source comes a report on well-conducted tests by Dr E. B. Philip on the relationship between experimental peak overpressure, p_0, and distance, R, for three UK Second World War bombs, shown in Figure 1.4. The bombs were 4000 lb high capacity with 72% charge weight, 1000 lb medium capacity with 44% charge weight, and 1000 lb General Purpose with 30% charge weight. The same filling, Amatol 60/40 was used for each bomb. Philip [1.15] proposed a useful, simple relationship between p_0 (in psi) and z as follows:

$$p_0 = \frac{a_1}{z} e^{b_1/z^{1/2}},$$ (1.9)

where, as defined above, $z = R/W^{1/3}$ in ft/lb$^{1/3}$ and typical values of a_1, and b_1 were:

$$
\begin{array}{lll}
\text{4000 lb bomb, high capacity,} & a_1 = 12.9, & b_1 = 5.7 \\
\text{1000 lb bomb, medium capacity,} & a_1 = 12.1, & b_1 = 5.7 \\
\text{1000 lb bomb, general purpose,} & a_1 = 11.4, & b_1 = 4.9
\end{array}
$$ (1.10)

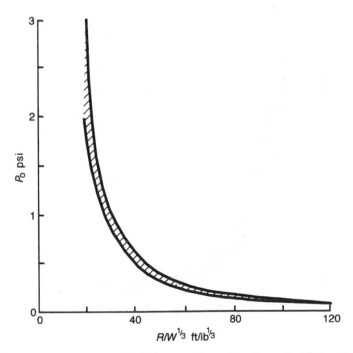

Figure 1.3 Peak overpressure v Scaled range for rectangular, cylindrical and spherical charges (from Stoner and Bleakney, ref. 1.13).

As Figure 1.4 shows, the differences were very small, and all the results lay within a narrow scatterband. Note that the results summarized in Figure 1.4 lie at the far right end of Philip's curve, because the charges were small and the peak pressures very low. Philip's work still remains as one of the most authoritative pieces of work on high-explosive bombs, although based on a dangerously small number of tests. Her results were also given by Christopherson in reference [1.9]. What they suggest is that the decay of peak instantaneous pressure with range is exponential, which seems logical. In fact the lower scatterband on Figure 1.4 is close to the 'inverse square law' given by $p_0 = 500/z^2$, which is a simple rule still in use in preliminary calculations (after conversion to metric units).

The 4000 lb high-capacity bomb tested by Philip must have been one of the 'blockbuster' bombs containing a high proportion of charge and a relatively low containment mass, designed by Barnes Wallis to attack heavily reinforced concrete protective structures by earth transmitted shock rather than by penetrative impact. It is interesting to note that during the Gulf War in 1991, when the standard 2000 lb laser-guided bombs dropped on bunkers by the US Air Force were clearly not entirely demolishing the targets, it was reported in the press and elsewhere that an accelerated manufacturing programme was undertaken to produce 4000 lb bombs. Readers who wish to see what a UK 4000 lb

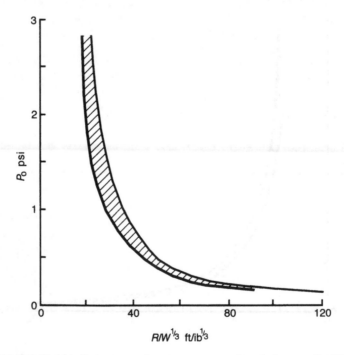

Figure 1.4 Relationship between peak overpressure and scaled range for UK Second World War bombs (from Philip, ref. 1.15).

bomb looked like may find examples mounted on plinths at certain RAF stations, but it is best not to stop! According to records, 68000 high capacity versions of this bomb were released over enemy territory between 1941 and 1945. Barnes Wallis has a further famous place in military history in connection with the design of bombs to destroy the main structure of dams, and with the development of the Tallboy and Grand Slam bombs (see section 7.3).

The characteristics of explosives are nowadays often presented dimensionally on a log log scale as a relationship between the peak instantaneous overpressure in bars and the distance in metres in any direction from the centre of the explosion of a spherical charge. Figure 1.5, due to Lavoie [1.16], gives a comparison for 1000 tons of TNT, FAE and a nuclear explosion. The peak overpressures close to the explosion vary considerably, as might be expected, but in all cases have diminished to 0.1 bar at a radius of 100 or 200 metres. Figure 1.6 from the same source compares impulses for similar explosions, and shows that, because of the duration of the positive phase, much more impulse is available from nuclear and FAE explosions than from TNT.

Some time after the end of the Second World War photographs of the first atomic explosion in New Mexico were released, and a year or two later, in 1949, a paper on the formation of a blast wave from a very intense explosion, written

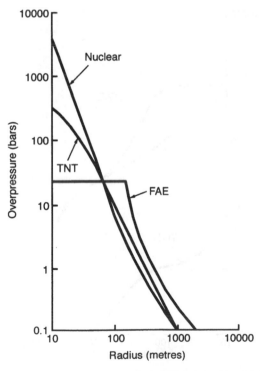

Figure 1.5 Overpressures for 1 KT explosions (from Lavoie, ref. 1.16).

by G. I. Taylor [1.17] in 1941 for the Civil Defence Research Committee of the Ministry of Home Security, was declassified. The predictions of this theoretical paper were confirmed by measurements of the radius of the luminous globe from the photographs, as it spread outwards from the centre of the explosion.

The theory indicated that the radius of the outward propagating spherical shock wave was linked to the time t since the beginning of the explosion by the equation

$$t = 0.926R^{5/2}\rho_a^{1/2}E^{-1/2}, \tag{1.11}$$

when the ratio of specific heats, $\gamma = 1.4$. As before, ρ_a was the atmospheric density and E the energy released by the explosion. There was doubt about the validity of the assumption that $\gamma = 1.4$ at all stages of the explosion, because at the extremely high temperatures of an atomic explosion γ could be increased due to dissociation and perhaps affected by intense radiation. However, when the photographs were examined, Taylor was able to compare the time t in milliseconds with the radius of the shockwave, and found that the relationship between $\frac{5}{2}\log_{10}R$ and $\log_{10}t$ was a straight line:

$$\tfrac{5}{2} \cdot \log_{10}R - \log_{10}t = 11.915. \tag{1.12}$$

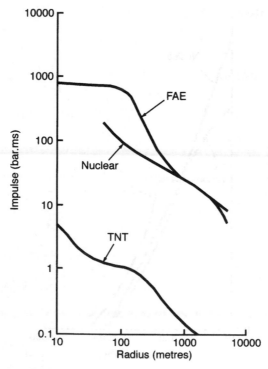

Figure 1.6 Impulses for 1 KT explosions (from Lavoie, ref. 1.16).

The assumption that $\gamma = 1.4$ was constant at all temperatures was vindicated. The explosion took place only 100 ft above ground, and Taylor pointed out that the fireball would have reached the ground in less than 1 msec; however, the photographs indicated that this impact did not affect the conditions in the upper half of the luminous globe.

As we noted in Eq. (1.1), the instantaneous maximum pressure at any distance R from a perfectly spherical charge is proportional to E/R^3, and Taylor gave the relationship as

$$p_0 = 0.155 \ E/R^3. \tag{1.13}$$

This raises an interesting fact. The peak pressure at radius R does not depend on atmospheric density, whereas the time t since the beginning of the explosion for the shock front to reach a radius R depends on $\rho_a^{1/2}$. Taylor used this to calculate the pressure-time relationship for a fixed point in terms of p/p_0, and t/t_0, where t_0 is the time for the shock front to reach that point.

By the mid-1950s the electronic computer had begun to transform the analytical work on blast pressures. It was possible to solve detonation problems without recourse to experimental data, because the hydrodynamic equations of motion which lead to non-linear partial differential equations could be integrated numerically in a relatively short time. Work in this field was summarized by Brode [1.18], who took a bare sphere of TNT of loading density 1.5 g/cm^3 and produced pressures, densities, temperatures and velocities as functions of time and radius. His relationship between p, the pressure in bars above atmospheric, and the shock radius parameter $R/W^{1/3}$ (where W is the weight of the charge at atmospheric pressure in kg and R is in m) is given in Figure 1.7.

A Group Technical Centre paper from ICI Explosives has shown how recent history has been influenced by the rapid increase in the availability of computer programs which can calculate the theoretical detonation characteristics of an explosive. Typical programs are TIGER, from Stanford Research International in the USA, and IDEX, from ICI Explosives in the UK. Other codes are available to help predict the practical performance in the field of explosives, depending on the conditions under which they are used, for example rock blasting.

Correct initiation can be developed by accurate modelling of initiation behaviour using finite element hydrocodes, such as DYNA, originally from the Lawrence Livermore Laboratory in the USA. Other codes have been written linking blast characteristics to fragmentation, vibration and the initiation sequence. A very simple assessment of the performance of explosives in practical circumstances, such as in boreholes, can be made by measuring the velocity of detonation. This can be found by high-speed photography or by a radar technique based on the use of microwaves. In general, the use of computers in all stages of the propagation of explosive blast has resulted in a rapid increase in scientific knowledge and engineering experience.

The nature of explosions

1.4 THE DECAY OF INSTANTANEOUS OVERPRESSURE IN FREE AIR

When Rankine, and later Hugoniot, analysed the pressure, velocity and density of a gas after the passage of a shock wave, they considered the conservation of mass, energy and momentum before and after the passage of the instantaneous jump in pressure. Their analyses are compared in ref. 2.38.

Suppose that in front of a shock wave travelling with velocity $\bar{u}$ the pressure, density and material velocity of the gas is p_1, ρ_1 and u_1, and that after its passage these quantities change to p_2, ρ_2 and u_2. Then, by conservation of mass,

$$\rho_2(\bar{u} - u_2) = \rho_2(\bar{u} - u_1), \tag{1.14}$$

and by conservation of momentum

$$p_1 - p_2 = \rho_2(\bar{u} - u_2)(u_1 - u_2). \tag{1.15}$$

Eliminating u_2 gives

$$\bar{u} - u_1 = \left[\frac{p_2 - p_1}{\rho_2 - \rho_1} \cdot \frac{\rho_2}{\rho_1} \right]^{1/2}, \tag{1.16}$$

which indicates that when, as is usual, the air in front of the shock wave is at rest, $u_1 = 0$ and the velocity of propagation $\bar{u}$ can be determined entirely in terms of the pressures and densities on either side of the discontinuity. As the

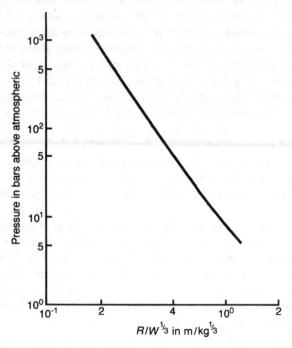

Figure 1.7 The prediction of Brode for a bare sphere of TNT of loading density 1.5 g/cm^3 (from Brode, ref. 1.18).

compression of the gas in the shock front is very fast, it is reasonable to assume that compression follows the adiabatic law (changes in pressure and volume with no change in absolute temperature). Then we may assume that the energy input is equal to the gain in energy in the gas, so that, by conservation of energy,

$$\tfrac{1}{2}(p_1 + p_2)(v_1 - v_2) = \frac{1}{\gamma - 1}(p_2 v_2 - p_1 v_1), \tag{1.17}$$

where v_1 and v_2 are the volumes occupied by unit mass, i.e. $v_1 = 1/\rho_1$, $v_2 = 1/\rho_2$. Eq. (1.17) has become known as the Rankine-Hugoniot equation.

The above analysis leads to the general relationship between shock front particle velocity, $\bar{u}$, and the speed of sound in air (u_a), where $u_a = (8p_a/\rho_a)^{1/2}$ and ρ_a is the density of the ambient air. This is

$$\frac{\bar{u}}{u_a} = \frac{5p}{7p_a}\left[1 + \frac{6p}{7p_a}\right]^{-1/2}, \tag{1.18}$$

where p_a is the pressure of the ambient air.

Also,

$$\frac{p}{\rho_a} = \frac{7 + 6p/p_a}{7 + p/p_a}, \tag{1.19}$$

where p is the density of the air behind the shock front. Taking u_a as 1117 ft/sec and p_a as 14.7 psi, Eq. (1.18) can be plotted as a relationship between $\bar{u}$ and peak overpressure (p_0) as shown in Figure 1.8.

The sudden discontinuous rise to p_0 is followed by a continuous decrease until the pressure returns to atmospheric and $p = 0$. The time between the arrival of p_0 and the return to atmospheric pressure is the 'positive duration' (see Figure 1.1), and for analytical purposes it is useful to represent the pressure-time curve as a mathematical function. Two functions are often used:

$$p = p_0(1 - t/t_0), \tag{1.20}$$

which is a simple triangular form, or more accurately,

$$p/p_0 = (1 - t/t_0)\, e^{-kt/t_0}, \tag{1.21}$$

where p is the pressure after any time t. By selecting a value for k (the wave form parameter), various decay characteristics can be indicated. Curves with very rapid decay characteristics are typical of nuclear explosions, and curves with slower decay rates are typical of explosions with large volumes of product gases. When $k = 1$, the positive and negative impulses (Figure 1.1) are equal, and the positive impulse is $p_0 t_0/e$. The curve given by Eq. (1.21) is often known as the Friedlander curve [1.19], because it comes from the work of F. G. Friedlander on behalf of the UK Home Office at the very beginning of the Second World War. Note that p/p_0 is non-dimensional and is therefore an intensity characteristic of the blast wave system.

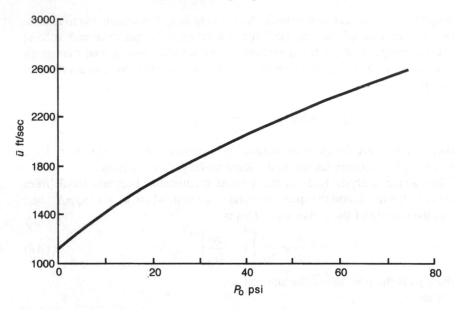

Figure 1.8 Relationship between shock front velocity and peak overpressure (from Rankine, ref. 1.3).

The duration of the positive phase, t_0, is a function of p_0 and the total energy yield of the explosion. The end of the positive phase has an overpressure of zero, which is the characteristic of a sound wave, since a sound wave has no shock front and only infinitesimal changes in pressure amplitude. So the zero over-pressure condition must move away from the centre of the explosion at the speed of sound in air. This is a lower velocity than the shock front velocity, $\bar{u}$, and means that the duration of the positive blast wave increases with distance, reaching a limiting value when $p_0 = 0$.

Typical values of t_0 for high explosives can be found from the formula

$$\frac{t_0}{W^{1/3}} = \frac{980(1 + (z/0.54)^{10})}{[1 + (z/0.02)^3][1 + (z/0.74)^6][1 + (z/6.9)^2]^{1/2}}, \qquad (1.22)$$

where t_0 is in milliseconds and W is in kilograms. As before, $z = R/W^{1/3}$ but in these units R is in metres.

For nuclear explosions,

$$\frac{t_0}{W^{1/3}} = \frac{180(1 + (z/100)^3)^{1/2}}{[1 + (z/40)]^{1/2}[1 + (z/285)^5]^{1/6}[1 + (z/50\,000)]^{1/6}}, \qquad (1.23)$$

where t_0 is now in seconds and W is in kilotonnes. The units of R are metres.

These equations are rather complicated, and we must turn again to the work of Dr Philip for a simpler formula. When she conducted the tests to compare p_0 and R, that led to Eq. (1.9), she also measured positive phase durations, and from these results proposed the relationship

$$t_0/W^{1/3} = c_1\,e^{-d_1/z^{1/2}},\qquad(1.24)$$

where $z = R/W^{1/3}$ in ft/lb$^{1/3}$, t_0 is measured in milliseconds, and typical values of c_1, and d_1, are:

$$4000\,\text{lb bomb, high capacity,}\qquad c_1 = 8.1,\quad d_1 = 4.8$$

$$1000\,\text{lb bomb, medium capacity,}\quad c_1 = 6.9,\quad d_1 = 4.2$$

$$1000\text{ lb bomb, general purpose,}\qquad c_1 = 6.2,\quad d_1 = 3.8$$

As Figure 1.9 shows, the scatter is small. Similar curves to Figure 1.9, but for nuclear explosions were given in many publications in the 1950s.

A knowledge of the instantaneous pressure and the duration of the positive phase allows us to calculate the blast impulse of an explosion. This is usually expressed per unit of projected area, and in imperial units is lb sec/in^2 or lb msec/in^2. For the three bombs analysed by Philip, she found that a simple relationship between impulse (I), charge weight and radius took the form

$$I = K_1 W^{2/3}/R,\qquad(1.25)$$

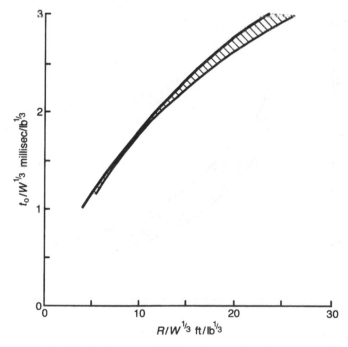

Figure 1.9 Relationship between scaled duration and scaled range (from Philip, ref. 1.15).

where I is in lb sec/in^2, W is in lb and R in feet. Values of K_1 were

$$4000\,\text{lb bomb, high capacity,} \qquad k_1 = 0.054,$$

$$1000\,\text{lb bomb, medium capacity,} \quad k_1 = 0.046,$$

$$1000 \text{ lb bomb, general purpose,} \quad k_1 = 0.038$$

Figure 1.10 shows the scatter of these results, which was noticeably greater than for pressure or duration.

In Kennedy's summary of Second World War test results [1.11], he also gave a logarithmic plot of positive impulse versus scaled distance for explosions of cast TNT in free air, as shown in Figure 1.11. This is probably the best data to use in preliminary design calculations, because Philip's results were not from explosions in clear air.

Christopherson [1.9] suggested that the values of K_1 given above by Philip were only appropriate for a range of $R/W^{1/3}$ exceeding 6, and that at lower values of $R/W^{1/3}$ the impulse given by Eq. (1.25) would be too high. He quotes a more pessimistic treatment by Kirkwood and Brinkley [1.20] which suggested that when $R/W^{1/3}$ in ft/lb$^{1/3}$ was equal to unity Eq. (1.25) gave values of I about three times too great, and about two times too great when $R/W^{1/3} = 3$. The true values probably lie somewhere between these extremes.

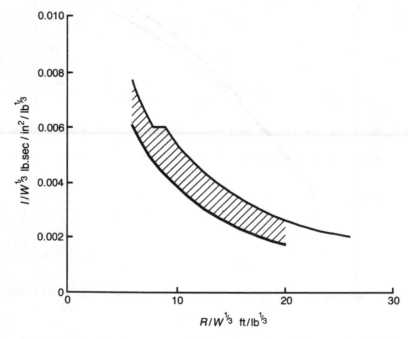

Figure 1.10 Relationship between scaled impulse and scaled range for UK Second World War bombs (from Philip, ref. 1.15).

The Kirkwood and Brinkley theoretical air-blast curves for cast TNT have become a much used standard for free-air explosions in a homogeneous atmosphere, and have been shown to be suitable for small explosions as well as nuclear explosions equivalent to 2 million pounds of TNT.

In a nuclear explosion the outward motion of the shock front when a burst occurs at, or just above, ground level is associated with hurricane-like winds that blow horizontally, parallel to the earth. The intensity of the wind is related to the peak overpressure, and rises from zero speed to a high value (perhaps 160 mph, or 256 km/hour) in almost zero time. This is quite different from a natural hurricane, in which wind speeds usually increase over a relatively long period. The dynamic pressure, q, due to the winds is given by

$$q = \tfrac{1}{2}\rho u^2, \qquad (1.26)$$

where ρ is the air density behind the shock front, and u is the velocity of the air particles in the shock front. Using the Hugoniot-Rankine equations it can be shown that

$$\frac{q}{p} = \frac{5}{2} \cdot \frac{p}{7p_a + p}, \qquad (1.27)$$

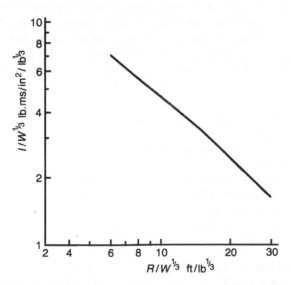

Figure 1.11 Relationship between scaled impulse and scaled range for cast TNT charges in free air (from Kennedy, ref. 1.11)

sometimes written in psi as

$$q = 14.7 \left[\frac{(5/14)(p/14.7)^2}{1 + (1/7)(p/14.7)} \right].$$
(1.28)

Note that the value of q is orders of magnitude greater than the external wind pressure used in the design of conventional surface structures. At 6 km from the centre of a 1 megaton explosion the peak overpressure would be 42 Kp_a and the windspeed 81 m/sec; the arrival time of the shock front would be 14 seconds and the duration of the positive pressure pulse 3.0 seconds. At the same distance from the centre of a 10 megaton explosion the figures would be 135, 200, 12.5 and 4.7. At 12 km the respective sets of figures would be: 1 megaton: 15, 31, 31, 3.7; and 10 megaton: 46, 96, 30 and 6.3.

Nuclear explosions also produce earth shocks in the neighbourhood of surface or underground structures. These shocks can arrive by direct travel through the earth from the point of detonation when this point is on or below the surface. Whether the air induced earth shock front outruns or lags behind the air blast wave front depends on the relationship between the air blast wave velocity $\bar{u}$, and the seismic velocity for the soil.

Direct earth shock travels at the seismic wave velocity (c_s), which varies with soil properties as shown in Table 1.1.

At depths of 100 feet or less, and for distances (R) for which $p_0 < 200$ psi, the direct earth shock produces low accelerations, whereas air-induced earth shock has much higher accelerations. Most protective structures are therefore designed against air-induced earth shock only.

Since the dynamic pressure (q) is proportional to the square of the wind velocity, it falls to zero some time later than the overpressure, and the dynamic positive phase is larger than the overpressure positive phase. By the time the wind stops blowing in a direction away from the centre of the explosion the overpressure is negative. There is consequently a reversal of wind direction, back towards the centre of the explosion, but at a relatively low velocity because

Table 1.1 Soil properties and seismic wave velocity

Material	(feet/sec)
Loose and dry soil	800–3300
Clay and wet soil	2500–6300
Coarse and compact soil	3000–8500
Sandstone	3000–14 000
Shale and marl	6000–17 500
Limestone – chalk	7000–21 000
Volcanic rock	10 000–22 600
Plutonic rock	13 000–25 000

it is caused by the negative phase.   As a concluding thought to this chapter, readers should be reminded that the formulae presented and the research described were mainly concerned with the behaviour of explosions at or near sea level. An examination of the dependence of blast analysis on ambient pressure and temperature was made in 1944 by R. G. Sachs at the Aberdeen Proving Ground, Maryland, USA, with particular reference to high altitude explosions. It was reported in reference [1.21].

1.5 REFERENCES

1.1   Earnshaw, S. (1858) On the mathematical theory of sound, *Phil. Trans.*, cl 133.
1.2   Lamb, H. (1895) *Hydrodynamics*, Cambridge.
1.3   Rankine, W. J. M. (1870) On the thermodynamic theory of waves of finite longitudinal disturbance, *Phil. Trans.*, cl 277, p. 530.
1.4   Hugoniot, H. (1887, 1889) Mémoire sur la propagation du mouvement dans les corps et specialement dans les gaz parfaits, *J. de l'ecole polytech.*, Paris, **57** (1887), and **58** (1889).
1.5   Davis, W. C. (1987) The detonation of explosives, *Scientific American*, **256**(5), May.
1.6   Taylor, G. I. (1939) *The Propagation and Decay of Blast Waves*, UK Home Office, ARP dept, RC 39, October.
1.7   Taylor, G. I. (1940) *Notes on the Dynamics of Shock Waves from Bar Explosive Charges*, UK Ministry of Home Security, Civil Defence Research Committee paper.
1.8   Taylor, G. I. (1941) *The Propagation of Blast Waves over the Ground*, UK Ministry of Home Security, Civil Defence Research Committee paper.
1.9   Christopherson, D. G. (1946) *Structural Defence, 1945*, UK Ministry of Home Security, Research and Experiments Department, RC 450.
1.10 *US Army Fundamentals of Protective Design (Non-nuclear)* (1965), Dept of Army Technical Manual TM5-855-1, Washington.
1.11 Kennedy, W. D. (1946) Explosions and explosives in air. In *Effects of Impact and Explosion*, Summary Tech. Rep. DW2, NRDC, Washington, Vol. 1, Chap. 2.
1.12 Lindberg, H. E. and Firth, R. O. (1967) Tech. Rep. AFWL-TR-66-163, Vol. 2, Air Force Weapons Lab, Kirtland, USA.
1.13 Stoner, R. G. and Bleakney, W. (1948) The attenuation of spherical shock waves in air, *Jour. Appl. Phys.*, **19**(7), 670.
1.14 Goodman, H. J. (1960) *Compiled Free-air Blast Data on Bare Spherical Pentolite*, BRL Report 1092, Aberdeen Proving Ground, Maryland, USA.
1.15 Philip, E. B. (1942) *Blast Pressure Time-Distance Data for Charges of TNT and GP Bombs*, UK Home Office Report REN 168.
1.16 Lavoie, L. (1989) Fuel-air explosives, weapons and effects, *Military Technology*, **9**.
1.17 Taylor, G. I. (1950) The formation of a blast wave by a very intense explosion, I theoretical discussion, *Proc. Roy. Soc.*, Series A, **201**(1065), March.
1.18 Brode, H. L. (1959) Blast wave from a spherical charge, *The Physics of Fluids*, **2**(2), March/April.
1.19 Friedlander, F. G. (1939) *Note on the Diffraction of Blast Waves by a Wall*, UK Home Office ARP Dept, RC(A) July; also (1940) *Diffraction of Blast Waves by an Infinite Wedge*, UK Ministry of Home Security, Civil Defence Research Committee report RC 61, February.

1.20 Kirkwood, J. G. and Brinkley, S. R. (1945) *Theory of Propagation of Shock Waves from Explosive Sources in Air and Water*, Div. 2, NRDC Rep. A 318, OSRD No. 4814, March.
1.21 Sachs, R. G. (1944) *The Dependence of Blast on Ambient Pressure and Temperature*, BRL report 466, Aberdeen Proving Ground, Maryland, USA.

2

The detonation of explosive charges

2.1 GROUND BURST EXPLOSIONS

When an explosive charge detonates in contact with an unyielding surface, in perfect conditions the shock wave in theory has a hemispherical wavefront, as shown in Figure 2.1, and not the spherical wave front of an explosion in free air. The energy of the explosion is therefore concentrated over a smaller total surface area, and theoretically all the relationships discussed so far for free-air bursts will still be valid if twice the charge weight (2 W) is substituted for W in the equations. However, if the earth is yielding and capable of absorbing part of the energy, the factor of 2 on charge weight must be judiciously reduced. Experiments suggest it is nearer 1.7, as discussed below.

Plenty of experimental data became available over the years for ground bursts of nuclear and conventional weapons, much more than was available for air bursts. This was partly due to the practical limits on the height above ground

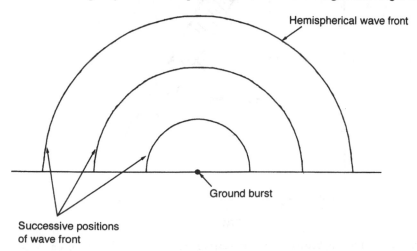

Figure 2.1 Hemispherical wave front for ground contact explosions.

that air bursts could be detonated (often the height of a tower). Most bursts were therefore subject to the effects of reflection. Tests during the Second World War suggested that the results for pressure and impulse were much more scattered than for air bursts, but nevertheless a most useful data collection was undertaken in 1946 by Kennedy [1.11]. This work was in connection with the decision of the US National Defence Research Council to produce an authoritative report on the *Effects of Impact and Explosion*, and Kennedy contributed the second chapter on 'Explosions and explosives in air'. This was a historically valuable report, although the data was obtained from tests with relatively undeveloped instrumentation.

Kennedy's summary of results for the pressure–distance relationship in ground burst tests is shown in Figure 2.2, taken from ref. [1.11], and his summary of the impulse–distance relationships is given in Figure 2.3, taken from the same reference. He proposed that the factor in Eq. (1.25) should be 0.054. He also proposed factors of 0.036 for heavily cased charges and 0.081 for bare charges. Data from ground bursts of hemispherical TNT charges between 5 and 500 tons in weight, and of nuclear charges ranging from 20 tons to 1.8 kilotons were later analysed by Kingery [2.1], and have been reported in detail by Baker [4.4]. Experimental results obtained at Suffield, Canada, in the 1960s on TNT hemispherical ground-burst charges in the 500 ton range have been

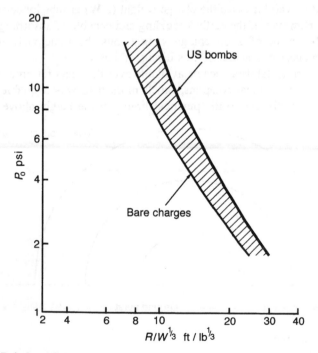

Figure 2.2 Relationship between incident pressure and scaled distance for Second World War bombs and bare charges (from Kennedy, ref. 1.11).

reported by Reisler *et al.* [2.2]. Typical results from all these tests are summarized in Figure 2.4. It was generally concluded that for ground bursts a reflection factor of 1.7 rather than 2.0 was appropriate, so the free-air burst equations can be used with 1.7 W substituted for W. The impulse–range relationship for the same series of tests is given in Figure 2.5.

Further test data for 20 ton spherical TNT charges, half buried in the ground, have been presented by Reisler *et al.* [2.3]. These results show close agreement with predictions based on the earlier analysis.

If the detonation of a high explosive occurs on or very near a flat surface, there will be a local impulsive load delivered to the surface in addition to the propagation of the blast wave through the air. This impulse, which is influenced by the size and shape of the charge, produces a shattering effect, or a 'brissance'.

The total impulse delivered to the flat surface was derived from test data, and is quoted in [1.10], where it is given as

$$I = W \left[183 - \frac{191}{1.1 + (ab/h^2)^{1/2}} \right] \text{lb sec/in}^2, \qquad (2.1)$$

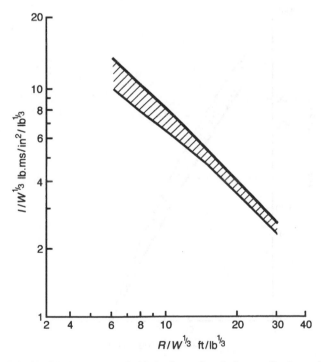

Figure 2.3 Relationship between scaled impulse and scaled range for Second World War bombs (from Kennedy, ref. 1.11).

where a and b are the dimensions of the charge in contact with the surface, and h is its height. The ratio ab/h^2 is often called the shape factor.

2.2 ABOVE-GROUND EXPLOSIONS

When an explosive source is located a set distance above a reflecting surface such as the ground, the reflection process is similar to that shown in Figure 2.6. When the incident wave first touches the flat surface it is reflected, but the reflected shock wave travels through the atmosphere at a higher velocity than the incident shock front. When it overtakes it, at what is known as the triple point, the fronts merge to form a single outward travelling front, known as the Mach stem, since it was first analysed by Mach and Sommer in 1877 [2.4]. A detailed review of the behaviour at the triple point was reported in 1959 by Sternberg [2.5], who studied the wide difference between theory and experiment for the shock wave angles at the triple point. The first solutions were given by von Neumann, in a paper on the oblique reflection of shocks [2.6], but there were difficulties in matching the range of incident wave angles at which Mach intersections are found.

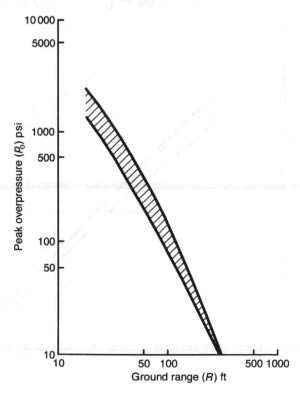

Figure 2.4 Peak overpressure v ground range for 20 ton hemispherical TNT charges (from Reisler *et al.*, ref. 2.2).

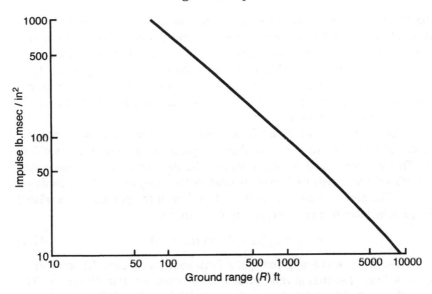

Figure 2.5 Impulse v. ground range for 20 ton hemispherical TNT charges (from Reisler *et al.*, ref. 2.2).

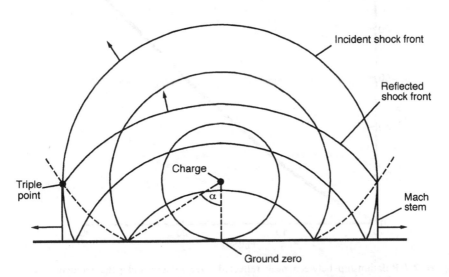

Figure 2.6 Reflection of shock waves for explosions above ground.

The Mach stem is not always a straight line, but is usually assumed to be so. As it travels outwards it acts in a similar way to the spherical shock front, in that there is an instantaneous rise in overpressure. The Mach stem is initiated when the angle of incidence (α) in Figure 2.6 exceeds 45°. In this figure we have shown the reflected waves as semi-circular in cross section, but this is not strictly correct because experiments show that the cross section more nearly approaches the shape of a semi-ellipse.

The reflected shock wave travels faster than the incident shock wave because the reflected overpressure is greater than the pressure existing in the incident wave. This is a very fundamental aspect of the interaction of pressure waves with surfaces which will be important later in the analysis of shock loads on structures. The peak overpressure at the flat surface of the ground, p_r, is related to the peak incident overpressure, p_0, by the formula

$$p_r/p_0 = 2(7p_a + 4p_0)/(7p_a + p_0). \tag{2.2}$$

This relationship was first given in the form shown in Figure 2.7 for a value of p_a of 14.7 psi. It applies at zero incidence, i.e. when $\alpha = 0$ in Figure 2.6. The formula does not change much as α is increased from 0 to 30°, but at higher values there is a noticeable change, as shown in Figure 2.8. It is intriguing that for low values of incident pressure (i.e. $p_0 = 5$ psi), the reflected pressure increases as the angle of incidence changes from 40° to 55°. This is because blast waves have finite amplitude, and the features of soundwave reflection (i.e. that incident and reflected waves have equal strengths) no longer apply.

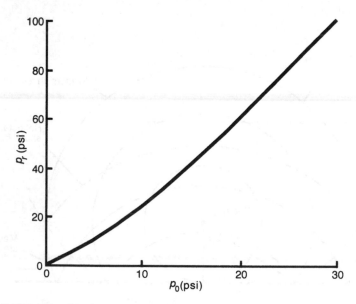

Figure 2.7 Relationship between peak reflected overpressure and peak incident overpressure when $p_a = 14.7$ psi.

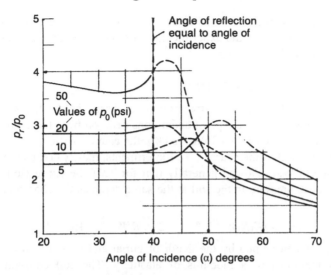

Figure 2.8 Variation of p_r/p_0 with angle of incidence (from Glasstone and Dolan, ref. 2.9).

The properties of obliquely reflecting shock waves have been a subject of much theoretical and experimental research over the years. Theories due to von Neumann [2.7] and more recently to Dewey and McMillin [2.8] have used different approaches (pseudo-stationary and non-stationary) to predict more accurately the point at which transition to Mach reflection occurs. It is near this transition that the total pressure exceeds the normal reflection pressure.

The theoretical results have been used to predict the peak hydrostatic over-pressure at ground level from an air burst explosion, and are often expressed as isobars of overpressure plotted on the axes of height of burst and ground radius. Curves of this type were presented by Glasstone and Dolan [2.9], and an adaptation of their results by Satori [2.10].

In a paper in 1985 by Dewey, Heilig and Reichenbach [2.11] on oblique reflections, experiments performed at the Ernst Mach Institut in Freiburg, Germany were reported for the first time. Small spherical charges of hexogen (0.016 to 1.024 kg) were detonated at heights ranging from 0.15 to 6.5 m over flat concrete reflecting surfaces. Pressure gauges were flush mounted in the surface between 0 and 8 m radius from a point immediately below the charge. The test results were compared with experiments at the Defence Research Establishment, Suffield, where 500 kg charges of TNT and 100 and 500 kg charges of pentolite were detonated at heights ranging between 1.2 and 44 m. The results were also compared with experiments at Suffield with 450 000 kg of TNT at a height of 4 m. By applying the scaling laws to the hexogen and pentolite explosions it was possible to compare the relation between p_0/p_a and the primary shock front radius in metres, on a basis of a 1 kg charge mass of TNT. This comparison is shown in Figure 2.9, where a small adjustment of 2% was made to allow for the energy release of hexogen as $1.02 \times$ TNT, although the pentolite results were not adjusted for the energy release ratio of $1.22 \times TNT$. The computed and experimental reflected shock pressure ratios are plotted as a zone in the same figure.

2.3 BELOW-GROUND EXPLOSIONS

The pressure in earth from the detonation of a charge below the surface is influenced by the distance of the point under consideration from the centre of the explosion (R), the weight of the explosive (W), the characteristic of the soil (K), and the depth of burial of the charge (H). Experiments, mostly conducted during the Second World War, showed generally that for TNT the expression for the instantaneous peak pressure in the soil at the same depth as the charge takes the form

$$p_0 = fK(R/W^{1/3})^{-n}, \qquad 2 < R/W^{1/3} < 15, \tag{2.3}$$

where $n = 3$, f is determined by the depth of burial and K is the soil constant having the dimensions of a modulus of elasticity. The soil constant varies tremendously, from 2000 psi (loam) to 100 000 psi (saturated clay) and 590 000 psi (limestone).

The form of Eq. (2.3) was proposed by Lampson [2.12] at the end of the Second World War, and the relationship between p_0 and $R/W^{1/3}$ for a silty clay

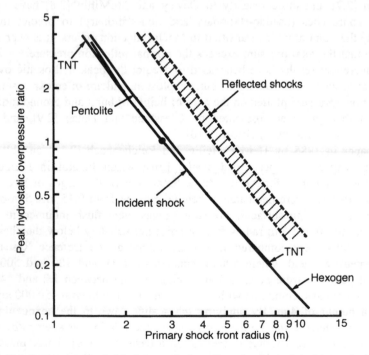

Figure 2.9 Pressure ratio behind a regularly reflected spherical incident shock front (from Dewey, Heileg and Reichenbach, ref. 2.11).

($K \simeq 5000$ psi) at the Camp Gruber location in Oklahoma, USA, is given by the heavy line in Figure 2.10. The figure also gives relationships for loam and unsaturated clay. Other functions for TNT can be constructed using the values for K listed in Table 2.1.

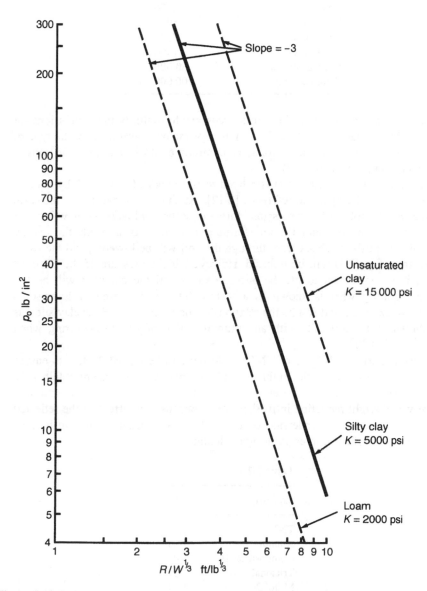

Figure 2.10 Relationship between peak overpressure and scaled range for explosions in soil (from Lampson, ref. 2.12).

Table 2.1

Soil type	K in Eq. (2.3) in lb/in^2
Dry topsoil	750
Moist topsoil	1500
Loam	2000
Silty clay	5000
Gravelly clay	7500
Sandy clay	1500
Compacted sand	50 000
Saturated clay	100 000

For explosives other than TNT it is necessary to multiply the expressions for p_0 by a further factor E, where E is an energy factor determined by the type of explosive. Lampson gave these for some common explosives of the time (mostly outdated now) (see Table 2.2).

The coefficient f in Eq. (2.3) is related to the depth of burial, H, by another curve given by Lampson in reference [2.12]. At shallow burial depths the earth above the 'chamber of compression' formed by the explosive gases will break up and be projected into the air, forming a crater. As a result the energy available to push the shock front through the soil will be lowered, and the value of p_0 from Eq. (2.3) will be reduced. However, if the pressure of the confined gases is not sufficient to lift the overburden of soil the chamber will be left intact, forming what is known as a 'camouflet'. At the time of Lampson's review of experimental results the relationship between f and H for clay silt was as shown in Figure 2.11, with an optimum value of $f(=1)$ occurring when $H/W^{1/3} = 2$ ft/lb$^{1/3}$.

Thus a charge of 1000 lb of TNT would have to be buried 20 feet to ensure the formation of a camouflet in this type of soil, whereas a charge of 500 lb TNT would need to be buried 16 feet. At depths of burial greater than the optimum there was a slight reduction in p_0 (i.e. $f > 1$) because the effect of the reflected shock wave from the underside of the soil surface began to diminish. This reduction was often ignored in design calculations.

Table 2.2

Explosive	E
TNT	1.00
Amatol	1.04
Composition B	1.04
Tritsonal	1.17
Minol 2	1.34
HBX2	1.30

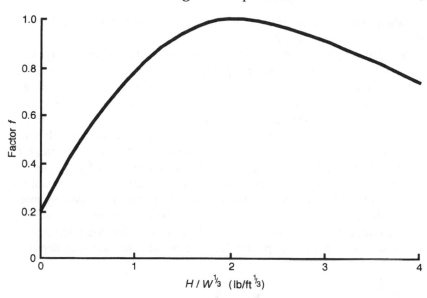

Figure 2.11 Relationship between the factor f and scaled cover depth for silty clay (from Lampson, ref. 2.12).

Lampson's work was still the accepted criteria for design in the 1960s, when the US Department of the Army Technical Manual TM5-855-1 on the fundamentals of protective design (non-nuclear) was published, and in the 1980s when further works on the design of protective structures were written. Very little fundamental work of an unclassified nature on this subject has been available to the general consulting engineer since the Second World War. There is also little experimental data available on the size of the chamber in a camouflet. Measurements in clay suggest that the chamber is usually spherical, with a volume of about 10 ft^3 per lb of TNT charge, and it is interesting to note that the products of combustion following an explosion, when expanded to the volume of the camouflet, would be approximately at atmospheric pressure.

A study on air blast from underground explosions as a function of charge burial was reported in 1968 by Vortman [2.13]. He was concerned with two constituents of an explosion set just below the surface of a soil mass, the first being the ground shock induced pulse which occurs when the ground shock is transmitted across the interface between the ground and the air, and the second from venting gases as the explosion erupts through a mound of rising earth. The peak instantaneous overpressure from venting gases would clearly reduce very much in magnitude as the burial depth increased, and by reviewing the results of many tests covering the period 1951 to 1965, Vortman was able to relate overpressure (psi) with scaled burial depth (ft/lb$^{1/3}$) at various scaled ground ranges. At a scaled ground range of 5 ft/lb$^{1/3}$ the relationship between ground

The detonation of explosive charges

shock induced peak overpressure (p_0 in psi) and scaled burial depth lay broadly within the band shown in Fig 2.12.

Much attention was given in the 1960s to the possibility of deep underground nuclear test explosions to avoid fall-out problems, occurring at such a depth that the only effects at ground level were associated with the propagation of elastic stress waves through the soil. To support the UK investigations a theoretical report on deep underground explosions was written by Chadwick, Cox and Hopkins [2.14]. One of the reasons for their study was that, although Hopkinson's size scaling law (which has been used extensively so far in this chapter) gives satisfactory results in most instances, there is a problem for scaling the crater size of very large nuclear explosions when the effects of gravity are no longer negligible. Thus craters from nuclear explosions become relatively more shallow as the explosion size increases. Further, the onset of plastic flow in a soil, as described by Coulomb's law of failure, means that the resistance to movement of a soil undergoing plastic deformation will increase with depth.

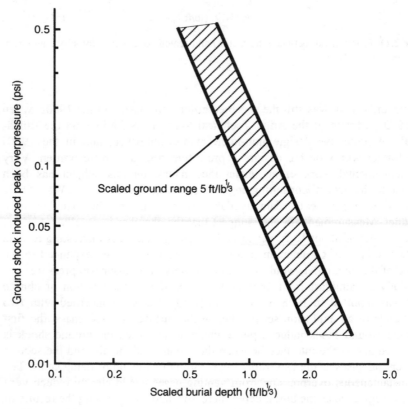

Figure 2.12 Experimental relationship between ground shock induced peak overpressure (p_0) and scaled burial depth, for a scaled ground range of 5 ft/lb$^{1/3}$ (from Vortman, ref. 2.13).

Table 2.3

Soil	Time to end of first expansion (ms)	Camouflet radius (feet)
Fully saturated clay	11.43	1.36
Partly saturated clay	7.47	1.02
Dry sand	7.47	1.02

Chadwick, Cox and Hopkins used a theory of spherical plastic-elastic flow in ideal soils to study the formation and size of the camouflet chamber in deep explosions and produced numerical values for the variation of camouflet radius with time during the first expansion phase for explosions at a depth of 100 feet for three idealized soils: fully saturated clays, partly saturated clays or mixed soils, and dry sands. The first of these has a cohesion >0 and an angle of internal friction =0; the second has values of both constants >0; the third has a cohesion =0 and angle of internal friction >0. The numerical results for the explosion of a 1 lb spherical charge (radius 0.137 ft) at a depth of burial of 100 ft were as given in Table 2.3 (figures are given for the end of the first expansion).

These results show that the main influence on camouflet radius is the angle of internal friction, because when $\phi > 0$ the figures coincide, irrespective of the cohesion. When $\phi = 0$ there is a noticeable increase in camouflet size.

There is a good description of the underground nuclear explosion in the book on the effect of nuclear weapons, compiled and edited by Glasstone and Dolan [2.9]. In their analysis of shallow explosions they point out that the ground surface moves upwards in the shape of a dome, which disintegrates and throws soil and rock fragments upwards. Much of this material falls back into the crater, entraining air and dust particles which are carried downwards and then outwards to produce a 'base surge' of dirt particles. It is carried downwind, and can occur several minutes after the original explosion. Meanwhile the 'main cloud' of the explosion, containing expanding gases and vapours, rises to a height of many thousands of feet.

They describe an underground nuclear explosion of 1.7 kilotons in 1957 at a depth of 790 feet. During the hydrodynamic phase of the explosion, when the shock wave is initiated and begins to expand outwards, the original chamber in which the bomb was placed expanded outwards to form a spherical cavity 62 feet in radius. Eventually this cavity could not support the overlying soil and the roof fell in to form a tall 'chimney' extending about 400 feet upwards from the burst point. Very roughly, the volume of the cavity was proportional to the cube root of the energy yield (assumed proportional to the equivalent weight of the charge in TNT). For contained explosions the constants for granite are 35 feet/$KT^{1/3}$, and for limestone 25 feet/$KT^{1/3}$. These figures apply to burst depths below 2000 feet. From the rough data assembled in this way it was concluded that in deep underground tests, where radioactive gases must not be allowed to escape, the depth of burial should not be less than 400 feet/$KT^{1/3}$, and should be increased even more in media with a substantial water content.

Before leaving the underground detonation of high explosives and nuclear weapons we must look briefly at three further aspects: positive impulse, the formation and size of craters, and the vertical and horizontal movements of the surface material near craters. The positive impulse per unit area of a pressure wave in soil due to a TNT explosion can be presented empirically in the form

$$I/W^{1/3} = 0.076\ K^{1/2}f(R/W^{1/3})^{-5/2}, \tag{2.4}$$

and the maximum positive impulse at a fixed target in the earth is three times as great as this. This maximum impulse is sometimes written as

$$I = 0.228\ K^{1/2}f(W^{1/2}/R^{2.5})\ \text{lb sec/in}^2. \tag{2.5}$$

The average duration of the positive pressure wave in soil, when the depth factor $f = 1$, is

$$t_0 = 0.174\ K^{-1/2}R^{2/5}W^{1/5}\ \text{sec}, \tag{2.6}$$

and the average duration of the pressure against a rigid target is

$$\bar{t}_0 = 0.368\ K^{-1/2}R^{2/5}W^{1/5}\ \text{sec}. \tag{2.7}$$

The fact that the pressure is maintained for a longer period against a fixed, rigid target than in free earth is a special feature of pressure waves in earth. It should be noted that the formulae given in Eqs (2.3), (2.4) and (2.5) refer to pressures measured at the same depth as the charge. Pressures measured at the same range, but at different depths, will not be as great. As an example, for a high explosive charge buried at the optimum depth $H = 2W^{1/3}$ ft, the pressure at the free soil surface will be 0.6 of that at the level of the charge.

When a charge explodes close enough to the surface to form a crater rather than a camouflet, the dimensions of interest are the crater diameter (D) and apparent depth h. The latter is the distance from the free surface to the uppermost surface of the crater debris, whereas H is the distance below the surface at which the charge was exploded. There are so many factors contributing to the final shape and size of the crater that there is considerable scatter in experimental results, and there is also a general shortage of experimental work in this field. Christopherson [1.9] reported on British tests carried out during the Second World War, and he began by classifying the crater types shown in Figure 2.13 for an explosion in clay.

In this diagram the type A crater is formed by an explosion on or very near the surface of the soil. The crater has a relatively clean surface, clear of rubble and debris. Type B crater, formed by a deeper explosion, contains much more debris that has fallen back and often has overhanging shoulders of the original soil. For a 250 kg charge in clay the depth of the explosion for type B would be about 10 feet. In type C, explosion depth 20 feet, the whole of the initial chamber of compression is filled with debris, which will not spread much beyond the limits of the crater. Christopherson made the point that most of the delay-fused bombs dropped in Britain during the Second World War penetrated to about 20 feet and produced type C craters. The deeper explosions (30 ft) produced a camouflet and a chimney, which we discussed earlier in relation to nuclear explosions, and deeper still (35 ft) the 250 kg charge in clay produced a spherical camouflet. These crater shapes were originally noted by Walley [2.15] following trials at Brancaster, where the soil was waterlogged blue clay. Later he surveyed earth shock effects in other types of soil, including the general classifications of clay, chalk and sandstone. The results are given in terms of the crater depth parameter $h/W^{1/3}$, the depth of the explosion parameter $H/W^{1/3}$, and also the crater diameter parameter $D/W^{1/3}$, in Figure 2.14.

Crater dimensions were also reviewed in Britain by Colonel F. W. Anderson in 1942 [2.16]. He suggested that the results could be predicted in different materials by introducing a soil factor to modify W for a given burial depth H. Taking clay as the basis (factor $=1.0$), he proposed factors of 2.75, 1.50 and 2.00 for chalk, soft sand and soft sandstone respectively. The British also looked at crater effects in 1945, when J. S. Arthur [2.17] produced a review of comparative performance between different types of high explosive. He also proposed an explosive factor to modify W for a given burial depth H. Taking TNT as the basis (factor $=1.0$) he suggested factors of 1.0 for Amatol, 1.20 for

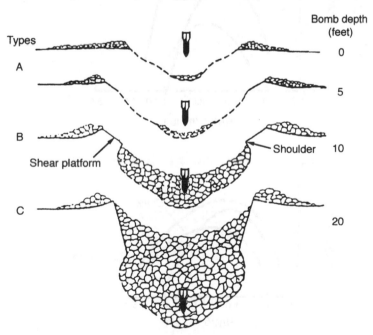

Figure 2.13 Types of crater for explosions in clay, reported by Christopherson (ref. 1.9) in 1945.

RDX/TNT (60/40) and 1.75 for aluminized explosives such as Torpex and Minol. At about this time an attempt was made by Devonshire and Mott [2.18] to relate camouflet volume and crater dimensions to the energy available in the explosion, and their work was later used to support the analysis of deep underground explosions by Chadwick, Cox and Hopkins that we discussed earlier.

Lampson included crater dimensions in his chapter on 'Explosions in earth' in reference [2.19], when he reviewed the wartime experiments for the National Research and Development Committee in the USA in 1946. His review included the British results, so it is not surprising that his proposed relationships between scaled crater dimensions and scaled depth of burial were

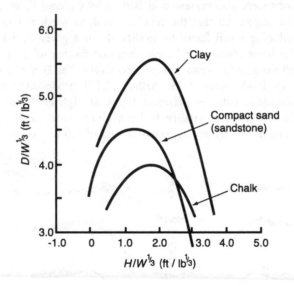

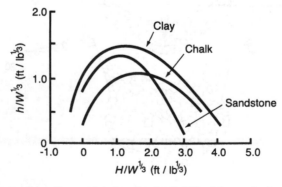

Figure 2.14 Crater dimensions related to the depth (*H*) of the explosion. Second World War tests reported by Walley (ref. 2.15).

very similar to those originated by Walley. Lampson's proposals are shown in Figure 2.15, and they later became incorporated in US military manuals, where they have remained relatively unchanged ever since. Lampson noted that crater diameter was sensitive to the charge depth, *H*, and less sensitive to the soil constant, *K*, and he therefore proposed that most of the proposed empirical curves for crater diameter could be approximately represented for TNT by the equation

$$D = 2.6fk^{1/2}W^{1/3}, \qquad \text{feet}(K \text{ in psi, } W \text{ in lb}). \qquad (2.8)$$

From Figure 2.11 *f* reached a maximum of unity when $H/W^{1/3} = 2.0$ ft/lb$^{1/3}$, and the curve of scaled crater diameter for clay also reaches a maximum at this value. The crater curves for chalk and sand, however, reach a maximum at a value of $H/W^{1/3}$ of about 1.5 and 1.0 respectively, which confirms that Figure 2.11 can only be strictly applied to clay soils.

When a crater is formed there are vertical and horizontal movements of the surface material in its vicinity. The surrounding material is violently displaced

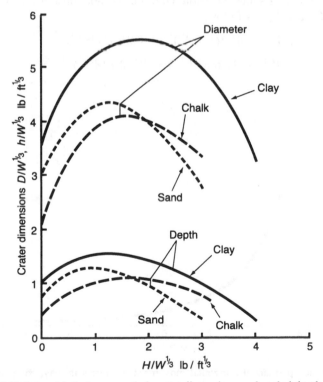

Figure 2.15 Relationship between scaled crater dimensions and scaled depth of burial of the charge (from Lampson, ref. 2.12).

outwards and upwards as the charge expands, and reaches a maximum displacement before some final contraction takes place. Final displacements, or permanent earth movements, are therefore slightly less than the maxima, but are much easier to measure by checking the positions of surface pegs before and after the explosion. Christopherson [1.9] reported observations in Britian for U_p, permanent horizontal movement and V_p, permanent vertical movement, as follows:

$$U_p/W^{1/3} = C_1(R/W^{1/3})^{-4}, \qquad (2.9)$$

where $C_1 = 200$ (clay), 25 (chalk) or 126 (soft sand), and

$$V_p/W^{1/3} = C_2(R/W^{1/3})^{-4}, \qquad (2.10)$$

where $C_2 = 145$ (clay), 19 (chalk), or 36 (soft sand).

The curve for the permanent horizontal movement in clay is shown in Figure 2.16. Note that at ranges greater than $R/W^{1/3} = 10$ ft/lb$^{1/3}$ there is virtually no horizontal movement at the surface. Also in clay and chalk the ratio between vertical and horizontal movement is about 1 : 1.4, whereas in soft sand it is about 1 : 4.

Tests in Oklahoma, USA, reported by Lampson, using dry sand as the medium, gave rise to a somewhat more complicated relationship for maximum movements (U_m and V_m) as follows:

$$U_m/W^{1/3} = 47.3(R/W^{1/3})^{-3} + 0.216(R/W^{1/3})^{-1}, \qquad (2.11)$$

$$V_m/W^{1/3} = 12.6(R/W^{1/3}) + 0.032(R/W^{1/3})^{-1}. \qquad (2.12)$$

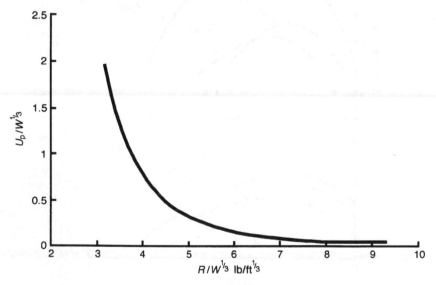

Figure 2.16 Scaled permanent horizontal movement of craters in clay, compared with scaled range. Information from research in UK during the Second World War, reported by Walley (ref. 2.15).

Table 2.4

Soil type	F
Hard rock	3.3
Soft rock	4.3
Hard chalk	4.7
Soft chalk	5.4
Blue clay	6.2
Loam	6.7
Gravel	6.8
Sand	7.5–7.8
Made ground	7.7–9.7

The distance from the centre of an underground explosion at which the face of a tunnel or trench would no longer suffer any disruption is known as the radius of rupture, R_r. Early British military handbooks on mining suggested that

$$R_r = F \cdot W^{1/3} \text{ feet,} \tag{2.13}$$

where the soil factor, F, is given in Table 2.4.

British tests during the Second World War suggested that the above figures were well on the safe side, and that the radius of rupture could safely be assumed to be 80% of that calculated from Eq. (2.13).

2.4 UNDERWATER EXPLOSIONS

The understanding of explosions that originate below the surface of the sea owes much to naval research and development of underwater weapons such as mines, depth charges and torpedoes. We are told that the first recorded uses of explosives to damage vessels at or below the waterline were at the siege of Antwerp in 1585 (Dutch v Spaniards) and at La Rochelle in 1625 (British v French), but these were essentially delayed-action floating bombs rather than true underwater devices. During the American Revolutionary War in the late eighteenth century mines were attached to ships below the waterline by a screw, and at the same time Bushnell developed a simple contact mine which was fired by a gunlock actuated when struck by the enemy ship. Further developments were recorded during the nineteenth century, and probably fused mines were used defensively during the Crimean War. The elementary fuses were glass tubes of sulphuric acid which were broken by impact with the hull of a ship. The acid mixed with chemicals to develop enough heat to cause the main charge of gunpowder to explode. By the First World War naval mines were laid in all parts of the world – perhaps a quarter of a million – and by the Second World War aerial mine laying had been perfected, as well as mine laying from submarines and surface craft like destroyers, motor launches, and specifically designed minelayers. The total mines laid by all combatants in the Second World War is thought to approach a half million. The total of merchant and naval vessels lost by all countries was probably approaching 2000 – a worldwide sinking rate of about one ship per 250 mines.

The powered torpedo, with a name derived from the family name for electric eels and rays, was first produced in its modern form by the English engineer Robert Whitehead in 1868. It had depth and lateral controls, propellers driven by compressed air, and a warhead of high explosive. Later the Americans developed a method of heating the compressed air by burning alcohol. This considerably raised the pressure available at the propellers and thereby increased the range. Torpedoes were originally fired underwater, but were later ejected from deck-mounted tubes and also launched from aircraft. By the end of the Second World War the total weight of a single torpedo was about 3000 lb.

Mines and torpedoes usually explode on or after contact with the target, but the other major naval weapon, the depth charge, was specifically designed to explode at greater depths on or near the seabed in the vicinity of a submarine. The design of depth charges therefore resulted in the scientific study of the nature of explosions deep below the surface. Work in this field later escalated with the coming of nuclear bombs and the need to assess the danger to naval vessels of underwater nuclear explosions some distance away.

Much of the fundamental work in Great Britain began when outstanding young scientists were mobilized to help the war effort in 1939 and 1940. Among them was W. G. Penney, later to become Lord Penney, and one of his first tasks was to examine the physics of underwater explosions on behalf of the Ministry of Home Security Civil Defence Research Committee. There had been some earlier work in this field at the end of the First World War, when H. W. Hilliar [2.20] produced a report on the properties of pressure waves from underwater explosions. He developed a piston gauge, similar in many ways to the crusher gauge we discussed in the Introduction, but with the piston and crusher initially at a set distance from the anvil, and not in contact with it. Hilliar found that the deformation of the crusher as a result of piston impact was a function of the energy of the piston alone, and using this he conducted 109 experiments to determine the major features of underwater shock waves. It is generally accepted that for its time Hilliar's work was of great simplicity and brilliance, since he was able to investigate the physics of shock waves, the principles of similarity and the weight/distance laws at the same time. A quarter of a century later his design of a multiple piston gauge was improved by G. K. Hartmann [2.21] in the USA. This work is described in detail in a good book on underwater explosions by R. H. Cole [2.22] first published in 1948. This monograph has proved to be one of the best written works on the subject, and includes the hydrodynamic analysis of the detonation process.

Returning to W. G. Penney, who died in 1991, he was assistant professor of mathematics at Imperial College, London, at the outbreak of the Second World War. Shortly after the outbreak of war he was loaned to the Ministry of Home Security and to the Admiralty to investigate the nature and properties of underwater blast waves. After a successful period of research he was sent to Los Alamos in 1944 to join the British team working on the development of the atomic bomb. He earned a high reputation there, and was chosen as one of the two British observers on the flight to Nagasaki, when the second atomic bomb was dropped. He returned to Britain to take responsibility for the design and development of British nuclear weapons and to become director of the UK Atomic Weapons Research Establishment. Under his leadership the first British atomic bomb was detonated in the Montebello Islands, off NW Australia, in 1952, and the first British hydrogen bomb was tested at Christmas Island, in the Pacific, in 1957. He was chairman of the UK Atomic Energy Commission between 1964 and 1967, and at the end of this period he became a life peer and returned to academic life as rector of Imperial College.

Penney [2.23] and Penney and Dasgupta [2.24] solved the differential equations for finite amplitude spherical underwater waves by a step-by-step procedure for TNT, using Riemann shock functions. The pressure rise in the shock wave is virtually instantaneous, as it is in air, and behind the shock wave there is an exponential fall in pressure, but this can never be less than the local pressure in the undisturbed water. There is therefore no negative phase. The explosive gases form a spherical bubble which rapidly expands until the pressure in the bubble is equal to the hydrostatic pressure of the water at the detonation depth. At this stage a violent contraction of the bubble takes place, followed by a second expansion, and so forth. Each expansion causes the spherical waves to propagate through the water in all directions. Because of the incompressibility of water the peak pressures are higher than would occur in air or in the earth, although the pressure decays more quickly.

Penney's first analysis made the assumption that changes from the initial detonation conditions were adiabatic, but this was amended in ref. [2.24] in the light of Taylor's work on spherical detonation waves [1.6]. For TNT with a loading density of 1.5 gm/cc the relationship could be represented empirically as

$$p_1 = 14100 \frac{W^{1/3}}{R} \cdot e^{0.27W^{1/3}/R}, \tag{2.14}$$

where p_1 is in lb/in^2, W in lb and R in feet.

Pressure wave analysis for underwater explosions was also undertaken in the early 1940s in the USA by J. G. Kirkwood in conjunction with H. A. Bethe [2.25], E. Montroll [2.26], J. M. Richardson [2.27] and S. R. Brinkley [2.28]. Their work was reported by the Office of Scientific Research in a series of publications. Starting from slightly different assumptions, Kirkwood and Brinkley [2.29] produced the formula

$$p_1 = 22\,150\,\frac{W^{1/3}}{R}\left(\log_{10}\frac{R}{W^{1/3}} + 0.873\right)^{-1/2}, \qquad (2.15)$$

for $R/W^{1/3} > 13$ ft/lb$^{1/3}$.

This gives good agreement with tests at the higher range of $R/W^{1/3}$, because the parameters of the theory were obtained directly from these tests. A comparison of calculated pressures for TNT has been made by Cole [2.22] and is shown in Figure 2.17. It was generally accepted that the Penney and Dasgupta analysis was more satisfactory for calculating the initial formation conditions for the shock wave. From Eq. (2.15) we can calculate impulse, which for $R/W^{1/3} > 13$ is given by

$$I = 2.11 W^{1/3}/R \qquad (I \text{ in lb m} \cdot \text{sec/in}^2). \qquad (2.16)$$

This equation is identical in form to that for explosions in air, see for example Eq. [1.25], but the underwater impulse for the same charge is about thirty times as large.

Kirkwood and Brinkley [2.29] calculated that the energy in the underwater shock wave diminishes rapidly as the wave progresses outwards, about 30% is dissipated within 5 radii of the charge and 48% within 25 radii. Energies found from experimental pressure/time curves are generally 25% lower than these calculations predict. The effect of substituting TNT with other types of explosive was examined, and generally it was found that the differences are not as

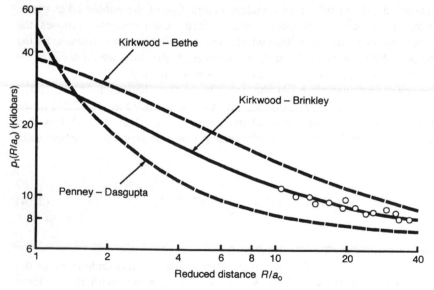

Figure 2.17 Comparison of calculated pressures for TNT (from Cole, ref. 2.22). R = distance travelled by shock wave; a_0 = original charge radius.

Table 2.5 Equivalent weights for explosives of the 1940s

Explosive	Equivalent weight of TNT (= 1.0) for blast effects	
	In water	*In air*
Torpex	1.22	1.36 (approx)
Minol	1.08	1.29
RDX/TNT 60/40	1.11	1.12
Amatol 60/40	0.87	0.83

great as in air or earth. Table 2.5 lists the equivalent weights for explosives of the 1940s given by Christopherson [1.9].

The first underwater nuclear test, code-named Baker, was carried out at the Bikini Atoll in 1946, when a nuclear bomb having a yield equivalent to 20 kilotons of TNT was detonated below the surface of the lagoon (200 feet deep). As the shock wave initiated by the expanding hot gas bubble reached the surface a rapidly expanding ring of darkened water was visible, followed by a white area as the shock wave was reflected by the surface. A column of water, the spray dome as it was called, was then formed over the point of burst. Its upward velocity was proportional to the peak pressure in the shock wave, and in the test it began to form 4 msec after the explosion. The upward velocity was initially 2500 ft/sec. A few milliseconds later the gas bubble reached the surface and a hollow column of gases vented through the spray dome. It is said in ref. [2.9] that the column was 6000 ft high with a maximum diameter of 2000 ft, and that 1 million tons of water were raised in the column. As the column fell back a condensation cloud of mist developed at its base and surged outwards. This bun-shaped cloud, known as the base surge, developed about 10 seconds after detonation, initially reaching a height of 900 ft. After about 4 minutes the height was double this, and the diameter about 3.5 miles. One of the first deep underwater explosions (in 2000 ft of water) took place in 1955, code-named Wigwam. During the pulsations of the gas bubble the water surrounding it had a large upward velocity and momentum, and the spray dome broke through the surface at a high speed. It is interesting to note that if the surface breakthrough occurs when the bubble is in the 'below ambient' phase, the spray column breaks up into jets that disintegrate into spray not unlike the vision of Moby Dick 'blowing'.

The analysis of the action of the gas bubble, whether from a nuclear or conventional underwater explosion, was a fruitful field of investigation during the 1940s. In 1941 Edgerton photographed the pulsations of the bubbles when detonator caps were exploded at various depths, and at about the same time Swift [2.30] and others filmed bubble motion when 0.55 lb of the explosive Tetryl was detonated 300 ft below the surface. The radius of the gas sphere for

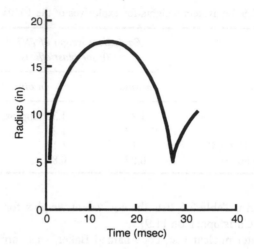

Figure 2.18 Radius/time curve of the gas sphere for 0.55 lb of Tetryl detonated 300 ft below the surface (from Swift, ref. 2.30).

this explosion is related to time in Figure 2.18, taken from the latter work. Note that the reversal of the bubble motion is virtually instantaneous and discontinuous. A comparison of radius and period, and the formulation of a simple analysis, had been made much earlier by Ramsauer [2.31] in 1923. He determined the position of the gas bubble boundary by means of an electrolytic probe, when charges of gun cotton weighing one or two kilograms were exploded at depths up to 30 feet in 40 feet of water. Although he could only measure the radius of the bubble and the time for isolated points, nevertheless it was a very useful and innovative study. He found that the variation of maximum bubble radius with depth could be predicted by the formula

$$\frac{4\pi}{3} p_0 a_m^3 \propto W, \tag{2.17}$$

where p_0 is the hydrostatic pressure at the depth of the explosion, W is the weight of explosive and a_m is the maximum bubble radius.

Photography, however, enabled a much more detailed picture of bubble behaviour to be drawn, and using Edgerton's results, Ewing and Crary [2.32], scientists from the US Woods Hole Oceanographic Institution, were able to plot curves of the type shown in Figure 2.19, and compare them with calculations by Herring [2.33] in 1941 based on the following formula for the first period of oscillation (T) of the bubble:

$$T = 1.83 a_m \left(\frac{p_0}{p_0}\right)^{1/2}, \tag{2.18}$$

where ρ_0 is the equilibrium density.

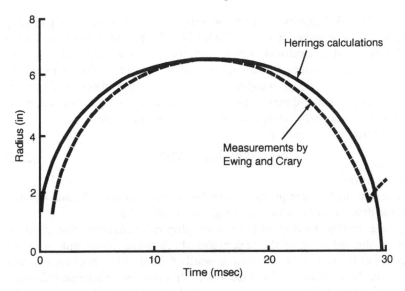

Figure 2.19 Measured and calculated radius of the gas sphere from a detonator one foot below the surface (from Herring, ref. 2.33).

Since the total energy associated with the radial flow of water (Y) is given approximately by

$$Y = \frac{4\pi}{3} p_0 a_m^3,$$ (2.19)

it is possible to express the first oscillation period in terms of total energy by

$$T = 1.14 p_0^{1/2} Y^{1/3} / p_0^{5/6}.$$ (2.20)

This is a formula attributed to H. F. Willis, a British scientist, and shows that the period is directly proportional to the cube root of the explosive charge weight, since for most high explosive the total energy varies with charge weight. The period also varies as $(d + 33)^{-5/6}$, where d is the water depth in feet, and this was verified experimentally by Ewing and Crary [2.32] who brought together the results of several studies to produce Figure 2.20. It was also shown that to a close approximation the periods of the first and second bubble oscillations are related by the formula

$$r_2 = r_1 \left(\frac{T_2}{T_1}\right)^2,$$ (2.21)

where T_1 and T_2 are the periods, and r_2 and r_1 are the fractional energies remaining after the first and second contractions.

For depths greater than 320 feet T_2/T_1 is very nearly constant and independent of depth, and is equal to about 0.6. This gives an energy ratio of 0.36.

The detonation of explosive charges

By 1942 the UK Ministry of Home Security had established a co-ordinating committee on shock waves, and G. I. Taylor [2.34] reported to this body on the vertical motion of a spherical bubble and the pressure surrounding it. He presented calculated values for the radius and the height of the centre of the bubble for a range of charge weights in a non-dimensional form. A. R. Bryant [2.35] took Taylor's non-dimensional analysis and produced a series of approximate formulae that involved simple calculations, for example:

Period of the first oscillation:

$$T = 4.32W^{1/3}/(d + 33)^{5/6}, \qquad (2.22)$$

(W in lb, d in feet)

His calculations for peak pressure p_0 in psi, duration in seconds, and distance R from the detonation point in feet are given in Table 2.6.

There are further fundamental areas of physical behaviour that must be mentioned: the reflection of the underwater shock wave downwards from the free surface of the water; the reflection upwards from the bottom of the sea, lake or river; and the generation of surface waves. In considering surface reflection, it is usually assumed that the explosion is sufficiently deep for the pressure pulse to behave as an intense sound pulse. This means that the depth is at least 12

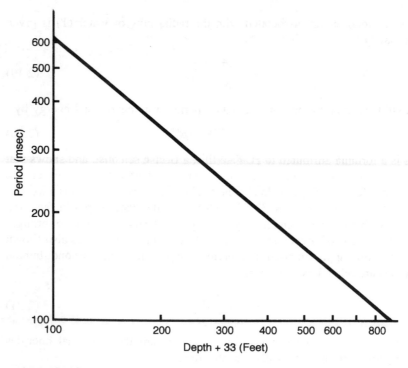

Figure 2.20 Bubble periods against depth for 0.55 lb Tetryl charges (from Ewing and Crary, ref. 2.32).

Table 2.6

Charge weight (lb)	Depth below sea level (ft)	$p_0 R$	Duration (sec)
800	150	11 450	0.113
	100	7090	0.148
	80	5350	0.169
256	150	10 600	0.078
	100	6800	0.101
	50	2950	0.151
16	80	4130	0.046
	50	2750	0.060
	30	1800	0.075
1	40	1750	0.026
	20	1160	0.034
	10	830	0.041

charge diameters. At the surface there will be a transmitted pulse (into air) and a reflected pulse which must be added to the original pulse. Because of the large difference in density and compressibility between air and water the transmitted pulse pressure will be very small and is usually neglected. The surface is then taken as undisturbed, which means that the reflected pulse must be equal and opposite to the initial pulse. This makes calculation of the effect of downwards reflection very simple, since at any point P it is a combination of a positive incident pressure at a distance r from the centre of the explosion and a negative rarefaction or 'tensile' pulse originating from an imaginary point which is at the image of the centre of detonation in the air immediately above this centre (i.e. the concept of images as used in the reflection of sound waves). The combination can theoretically produce a sharp decrease or 'cut-off' in the shock at point P, although in practice the decrease is more gradual.

Reflection of a pressure pulse from the seabed can vary considerably, depending on the state of the bed surface. On a rocky bottom the main pulse may give rise to a reflected wave having a peak pressure over half that in the main pulse; on a soft mud base the reflected pressure could be almost negligible. In general, the effect of the seabed is positive, leading to additional pressure. An upper limit to this additional pressure can be found by assuming that complete reflection occurs and that the concept of images can be used to calculate total pressure at a given distance from the centre of detonation.

Some examples were given by Cole in ref. [2.22]. Apparently the peak pressure, impulse and energy flux density 60 ft from a 300 lb TNT charge fired on a bottom of hard-packed sandy mud, was increased by 10, 23 and 47% over the values observed from a charge at mid-depth. These increases correspond to increasing the weight of a charge in free water by between 35 and 50%, rather than by a factor of 2, so a good deal of shock wave energy was transmitted and dispersed on the bottom.

Surface waves can be generated by underwater explosions or by air bursts close to the water surface. In the case of underwater explosions the outward propagating waves result from the gas bubble breaking through the surface, and in deep water explosions it is usually assumed that between 2% and 5% of the yield of the charge is directed to the energy of the train of surface waves. When the detonation is in relatively shallow water, there is often only an initial, solitary wave, particularly for very large nuclear explosions. After the 20 kiloton shallow water explosion at Bikini, mentioned earlier, wave heights and times of arrival were recorded at increasing distances from the detonation point (surface zero). At distances of 330, 2000 and 4000 yards the arrival times were 11, 74 and 154 seconds respectively, and the wave heights were 94, 16 and 9 feet.

Characteristic properties of the underwater shock wave from nuclear explosions have been given by Glasstone and Dolan [2.9]. For water with no reflections or refractions the relationship between peak instantaneous underwater shock pressure in psi and the slant range in thousands of yards is given in Figure 2.21. Approximate relationships have been established for surface wave amplitude (H) and radius (R) from surface zero for nuclear explosions as follows:

In deep water, $850\,W^{1/4} > d > 256\,W^{1/4}$, where W is yield in kilotons of TNT equivalent, and d is water depth in feet,

$$H = 40\,500\,W^{.54}/R, \tag{2.23}$$

where H and R are measured in feet.

In shallow water, $d < 100\,W^{1/4}$,

$$H = 150\,dW^{.25}/R. \tag{2.24}$$

The creation of surface waves by air bursts close to the surface had considerable military significance, and much of the research and testing during the 1960s was classified and not available through open publication. This is still thought to be the case. However, a paper by Kranzer and Keller [2.36] in 1959 was published in the *Journal of Applied Physics*, and their theory is now generally accepted as predicting wave train amplitudes for detonations of known size and position. They applied their general analysis of the distribution of impulse acting on the surface, and of the depression or elevation of the surface under explosive impulse. They suggested that the wave pattern contains a single maximum which moves outwards with a velocity $0.42\,(gH)^{1/2}$, where H is the height of the explosion above the surface. The wavelength at this maximum is $4.4\,H$, and the period is $5.1\,(H/g)^{1/2}$. The analysis seems to be invalid when $H = 0$, i.e. the explosion takes place on the surface, for it gives values of velocity and wavelength of zero. This is clearly not true. The authors discuss this and admit that the analysis is only suitable for larger values of H.

The work of Cole [2.22], which has been so useful in making this survey, has been summarized in more recent years by Kaye [2.37], and there is a useful

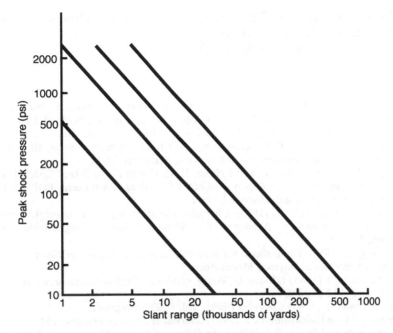

Figure 2.21 Relationship between peak instantaneous underwater shock pressure and the slant range for underwater nuclear explosions (from Glasstone and Dolan, ref. 2.9).

review of surface and seabed interactions in a book by Smith and Hetherington [2.38] of Cranfield University (Royal Military College of Science). Researchers might also usefully consult the work of Henrych [2.39], mentioned in the Introduction to this book.

2.5 EQUIVALENT PRESSURE FACTORS

We have already noted the custom of using TNT as the basis for information on blast pressure. This information can be applied to other explosives by multiplying the mass of the explosive by a conversion factor, as described in Tables 2.2 and 2.5.

Equivalent pressure factors for well-known modern explosives are 1.27 (PETN), 1.40 (Pentolite) and 1.07 (Tetryl). A recently developed explosive that became famous as the result of terrorist activity is Semtex, manufactured in what is now the Czech Republic. Some information on its behaviour is given in a paper by Makovicka [2.40], which included a figure giving the relationship between peak incident pressure and the weight of charge at a point 7 metres distant from the centre of the explosion. This relationship can be expressed approximately as follows:

Weight of charge (grams)	300	600	800	1000	1200
Peak incident pressure at 7 m (KPa)	15	22	25	27.5	28

A 1000 gram charge of TNT at 7 m, would give a peak incident pressure of about 20 KPa, according to parameters summarized recently by Kingery and Bulmash [2.41] and quoted in reference [2.38].

2.6 REFERENCES

2.1 Kingery, C. N. (1968) *Parametric Analysis of Sub-kiloton Nuclear and High Explosive Blast*, BRL Report 1393, Aberdeen Proving Ground, Maryland, USA.

2.2 Reisler, R. E., Keefer, J. H. and Griglio-Tos, L. (1966) *Basic Air Blast Measurements from a 500 Ton TNT Detonation*, Project 1.1, Operation Snowball, BRL memo report 1818, Aberdeen Proving Ground, Maryland, USA.

2.3 Reisler, R. E., Griglio-Tos, L. and Kellner, R. C. (1966) *Ferris Wheel Series*, Flat Top event, Project Offices Report, Project 1.1, Airblast Phenomena POR-3001, Aberdeen Proving Ground, Maryland, USA.

2.4 Mach, E. and Sommer, J. (1877) *Uber die Fortpflanzungsgesch windigkeit von Explosions schallwellen*, Akademie der Wissenschaften, Sitzangberichte der Wiener, Vol. 74.

2.5 Sternberg, J. (1959) Triple shock wave intersections, *The Physics of Fluids*, 2(2), American Institute of Physics, March/April.

2.6 von Neumann, J. (1943) *Oblique Reflection of Shocks*, Explosives research rep. No. 12, Buord, US Navy Dept.

2.7 von Neumann, J. (1943) *Collected Works*, Vol. 6, 239, Pergamon.

2.8 Dewey, J. M. and McMillin, D. J. (1985) *Journal of Fluid Mechanics*, **152**, 67.

2.9 Glasstone, S. and Dolan, P. J. (1962) *The Effects of Nuclear Weapons*, US Depts of Defense and Energy, Washington.

2.10 Satori, L. (1983) *Physics Today*, March.

2.11 Dewey, J. M., Heilig, W. and Reichenbach, H. (1985) Height of burst results from small scale explosions, *Proceedings (11) of 9th International Symposium on Military Applications of Blast Simulation*, Oxford, England, September.

2.12 Lampson, C. W. (1946) Effects of impact and explosions, *Explosions in Earth*, NRDC Washington, USA, Vol. 1, Chapter 3.

2.13 Vortman, L. J. (1968) *AirBlast from Underground Explosions as a Function of Charge Burial, Prevention of and Protection against Accidental Explosion of Munitions, Fuels and Other Hazardous Mixtures*, New York Academy of Sciences, ed. E. Cohen, Vol. 152, Art. 1.

2.14 Chadwick, P., Cox, A. D. and Hopkins, M. G. (1964) Mechanics of deep underground explosions, *Phil. Trans. Roy. Soc.*, Series A, No. 1069, Vol. 256, April.

2.15 Walley, F. (1944) *Note on Water Formation in Puddle Clay, Brancaster Beach*, UK Home Office Research Report REN 318, January.

2.16 Anderson, F. W. (1942) *Crater Dimensions from Experimental Data*, UK Ministry of Home Security, Civil Defence Research Committee, Report RC 344, September.

2.17 Arthur, J. S. (1945) *The Comparative Performance of Various Bomb Fillings – Crater and Earthshock Effects*, UK Home Office Research Report REN 520, June.

2.18 Devonshire, A. F. and Mott, N. F. (1944) *Mechanism of Crater Formation*, Theoretical Research Report No. 26/44, UK Armament Research Dept (AC 6995), July.

2.19 Lampson, C. W. (1946) Effects of impact and explosion, *Explosions in Earth*, Summary Tech. Dept. DW2, NRDC Washington, USA, Vol. 1, Chapter 3.

2.20 Hilliar, H. W. (1919) *Experiments on the Pressure Wave Thrown out by Explosions*, UK Admiralty Experimental Station Report.

2.21 Hartmann, G. K. (1946) *Taylor Model Basin (TMB), Report*, 531.

2.22 Cole, R. H. (1965) *Underwater Explosions*, Princeton University Press 1948, and Dover Publications, New York.

2.23 Penney, W. G. (1940) *The Pressure-Time Curve for Underwater Explosions*, UK Ministry of Home Security, Civil Defence Research Committee Report RC 142.

2.24 Penney, W. G. and Dasgupta, H. K. (1942) *Pressure-Time Curves for Submarine Explosions* (2nd paper), UK Ministry of Home Security, Civil Defence Research Committee Report RC 333.

2.25 Kirkwood, J. G. and Bethe, H. A. (1942) *The Pressure Wave Produced by an Underwater Explosion – Basic Propagation Theory*, US OSRD Report 588.

2.26 Kirkwood, J. G. and Montroll, E. W. (1942) *The Pressure Wave Produced by an Underwater Explosion – Properties of Pure Water at a Shock Front*, US OSRD Report 676.

2.27 Kirkwood, J. G. and Richardson, J. M. (1942) *The Pressure Wave Produced by an Underwater Explosion – Properties of Salt Water at a Shock Front*, US OSRD Report 813.

2.28 Kirkwood, J. G., Brinkley, S. R. and Richardson, J. M. (1953) *The Pressure Wave Produced by an Underwater Explosion – Calculations for Thirty Explosives*, US OSRD Report 2022.

2.29 Kirkwood, J. G. and Brinkley, S. R. (1945) *Theory of the Propagation of Shock Waves from Explosive Sources in Air and Water*, US OSRD Report 4814.

2.30 Swift, E. *et al.* (no date) *Photography of Underwater Explosions*, US Navy Bureau of Ordnance.

2.31 Ramsauer, C. (1923) *Ann. d. Phys.*, **4**(72), 265.

2.32 Ewing, M. and Crary, A. (1941) *Multiple Impulses from Underwater Explosions*, Woods Hole Oceanographic Institution report.

2.33 Herring (1941) *Theory of the Pulsations of the Gas Bubble Produced by an Underwater Explosion*, US NRDC Division 6 Report C4-Sr20.

2.34 Taylor, G. I. (1963) The vertical motion of a spherical bubble and the pressure surrounding it. In *The scientific papers of Sir Geoffrey Ingram Taylor*, Vol. 3, Cambridge University Press, 320.

2.35 Bryant, A. R. (1950) The behaviour of an underwater explosion bubble: further approximations, *Underwater Explosion Research*, Vol. 2, The gas globe office of Naval Research, US Dept of Navy.

2.36 Kranzer, H. C. and Keller, J. B. (1959) Water waves produced by explosions, *Journal of Applied Physics*, **30**(3), March, 398.

2.37 Kaye, S. H. (1983) *Encyclopaedia of Explosions and Related Terms*, US R and D Command, Large Calibre Weapons Systems Laboratory, Technical Report.

2.38 Smith, P. D. and Hetherington, J. G. (1994) *Blast and Ballistic Loading of Structures*, Butterworth Heinemann, Oxford.

2.39 Henrych, J. (1979) *The Dynamics of Explosion and Its Use*, Elsevier, Amsterdam (translation).

2.40 Makovicka, D. (1994) Influence of short shock load on response of masonry structure. In P. S. Bulson (ed.), *Structures Under Shock and Impact* (3), Computational Mechanics Publications, 53.

2.41 Kingery, C. N. and Bulmash, G. (1984) *Airblast Parameters from TNT Spherical Air Burst and Hemispherical Surface Burst*, Tech. Report ARBRL-TR-02555, US Army Armament Research and Development Center, Ballistics Research Lab., Aberdeen Proving Ground, Maryland, USA.

3

Propellant, dust, gas and vapour

3.1 LIQUID PROPELLANT EXPLOSIONS

We now turn to the nature of explosions other than those caused by TNT or nuclear-type detonation, and the first of these is the accidental explosion of the liquid propellant used in missiles and space rockets. Fully fuelled missiles have been known to fall back onto the launching pad and explode, forming a fireball and causing major structural damage, and it is important to know the factors involved in order to assess structural response. The characteristics of liquid propellant explosions were discussed by Fletcher [3.1], who was a member of the US Manned Spacecraft Center at Houston in the mid-1960s, and much of what follows is taken from his contribution to the New York Academy of Sciences conference on the prevention of the accidental explosion of hazardous mixtures, held in 1968. He pointed out that liquid propellant explosions occur in two phases, detonation followed by deflagration. The detonation process is limited, soon becomes extinguished, and is rather unsteady. This is in contrast to the detonation of TNT, which as we have seen proceeds through the mass of explosive in an orderly way. The detonation of liquid propellant continues for a longer period and at a lower pressure than that of an equivalent quantity of conventional explosive.

Detonation only occurs in a relatively small proportion of the total mass of the propellant, and most of the mass is consumed by deflagration, during which a large amount of chemical energy is transformed into thermal energy. The fireball is spherical and grows to a diameter of several hundreds of feet in a few seconds. The diameter in feet is about 10 times the cube root of the propellant weight in lbs, and the duration about 0.2 times the cube root of weight. The fireball growth from the contents of an Atlas missile reaches a maximum of about 400 feet diameter in approximately 1.5 seconds.

The shock wave from the detonation gives pressures that decrease linearly from the instantaneous peak, in contrast to TNT shocks, which decay exponentially. Close to the centre of the explosion peak propellant pressures are less than half those of an equivalent TNT explosion, but at greater distances the peaks are more nearly equal.

A study of the fireball from an exploding rocket was given by High [3.2] in 1968, and he collected experimental results relating diameter and duration to total propellant weight in lbs; his empirical relationships, derived by a least squares regression analysis of the data, were

$$D = 9.82W^{0.32},\qquad(3.1)$$

where D is diameter in feet, and W is propellant weight in lbs, and

$$T = 0.232W^{0.32},\qquad(3.2)$$

where T is duration in seconds. It was from these relationships that Fletcher's more approximate rules were derived.

The velocity with which the fireball lifted off from the ground was found to be relatively independent of weight. An exploding Saturn V rocket fireball was found to lift off when the radius reached about 700 feet, 11 seconds after detonation, with a velocity of 47 feet/sec.

3.2 DUST EXPLOSIONS

Combustible dusts of coal and peat form a world-wide hazard in mines, industry and at power plants. Explosions of grain dust, particularly in the mid-west of the USA, have resulted in considerable loss of life and structural disasters in grain elevator buildings. It is therefore of considerable importance that engineers should be able to calculate the blast characteristics of this type of deflagration or detonation.

The history of dust explosions is interesting. According to Chiotti [3.3] one of the first-known dust cloud explosions took place in a flour mill in Turin, Italy, in 1785. It was a hundred years before it was realized that ignition of the dust (as opposed to ignition of a flammable gas given off by the dust) could have initiated the explosion. The research which led to this understanding came about because of the great loss of life in coal mines during the nineteenth and twentieth centuries. We are told that in 1907 1148 miners were killed in the USA, and a considerable proportion died as the result of explosions. Between 1878 and 1913 there were 72 dust explosions and 60 deaths in the agricultural industries of the USA, and it is interesting that in all recorded industrial dust explosions in that country grain elevators were the setting for maximum occurrences, maximum injury and maximum damage to property. From 1925 to 1956 there were five times as many dust explosions in grain elevators than in flour mills or starch industries, and 1.5 times as many as in feed or cereal mills. In 1921 the largest grain elevator in the world, in Chicago, was destroyed by a dust explosion, even though considerable efforts had been made in the design to make the building fireproof by removing ignition hazards. As recently as the period 1958 to 1975 there have been over 20 grain elevator dust explosions in

Nebraska and Iowa, and over five feed mill dust explosions in Iowa and New York. The number of explosions related to total storage capacity during this period was greatest in Tennessee, Alabama and Nebraska. What is ominous is that in over 60% of recorded grain elevator and feed mill dust explosions the location and ignition source of the primary explosion remains unknown.

Grain dust and coal dust sustain similarly sized particles, in the size range between 1 and 150 micron. The determination of representative linear dimensions is not easy, because most dust particles are irregular in shape. In a grain elevator there are four types of dust: dump pit dust, belt-loading dust, main elevator dust and bean dust. About 95% of dump pit dust consists of bees' wings, the belt loading dust is mainly starch dust, and the main elevator dust is a 60:40 mix of bees' wings and starch. All these dusts are light, fine and low in ash; they prefer to stay in suspension and are easy to inflammate. The presence of so many bees' wings is intriguing. According to Matkovic [3.4] they are flat and round and difficult to control. They wander!

Research into the physics of dust explosions has concentrated on experimental work at laboratory scale and some attempt at theoretical predictions. Tanaka [3.5] suggests that there are three fundamental conditions to be investigated, the ignition point, the minimum concentration of dust at which an explosion can occur, and the flame propagation velocity of the explosion. One of the first analytical studies of ignition was made in 1936 by Parker and Hottel [3.6], and more recently, in 1959, Cassel and Liebman [3.7] considered a spherical space of dust cloud in which the particles were uniformly distributed. The rate of heat generation from the cloud is a function of the rate of oxidation of the constituent particles, and the rate of heat leaving the cloud is a function of the coefficient of transfer from solid to gas, the thermal conductivity of air, particle size and ignition temperature. There appears to be a most dangerous particle size giving the lowest ignition temperature, as shown in Figure 3.1 which relates ignition temperature with particle diameter for coal dust experimentally and theoretically. The experimental data is due to Kurosawa [3.8].

The minimum requirement for a flame to propagate is that the burning out of the particle occurs at the same time as the ignition of the next particle to it. Using this as a basis the minimum explosive limit concentration can be calculated as a function of particle diameter. Computation and experiments show that for clouds of plastic particles such as ethyl cellulose or cellulose acetate, the relationship between concentration in milligrams per litre and diameter in centimetres is approximately linear, and at a particle diameter of 0.05 cm the minimum explosive limit concentration is between 400 and 500 mg/litre. Flame propagation velocity appears to approach a maximum of 40 to 50 m/sec for a number of materials, reaching this value at a time of between 0.01 and 0.03 seconds from initiation.

Dust cloud explosions, particularly in agricultural products, have been a problem in Japan. A great deal of grain is imported, mainly from the USA, and stored in grain elevators at sea ports. Between 0.1% and 1% of bulk grain is in

Propellant, dust, gas and vapour

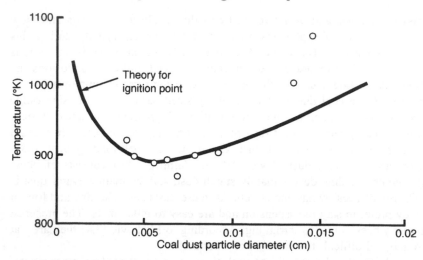

Figure 3.1 Comparison of theoretical and experimental ignition points for coal dust (from Kurosawa, ref. 3.8).

the form of powder, which forms a suspended dust with an explosive strength much higher than that of coal dust. Between 1962 and 1975 there were 23 agricultural dust explosions, killing 8 people and injuring 49. The main cause was sparks from welding processes in bucket elevators, chutes, bins and pipes. Because of this a number of university experimental research programmes were undertaken, using apparatus to generate uniform dust clouds over a wide range of concentrations. In one of these studies, reported by Enomoto [3.9], the apparatus consisted of a cylindrical explosion chamber, diameter 270 mm, 10 litres volume, that could be rotated to make a uniform dust cloud within it. Ignition at the centre of the chamber was by 0.1 gram of guncotton, detonated electrically. It was found that a maximum explosion pressure of about 6.5 kg/cm² occurred at a dust concentration of approximately 1000 g/m³ for cornstarch, potato starch and wheat flour. Maize and wheat dust produced explosion pressures between 4.5 and 5.0 kg/cm² over a range of concentrations between 1000 and 4000 g/m³. The rate of pressure rise for wheat dust reached a maximum of 60 kg/cm²/sec at a concentration of about 1500 g/m³, but the average rise was about half this. The relationship between the maximum rate of pressure rise $(dp/dt)_m$ and the ratio of the peak pressure (p_m) and initial pressure (p_i) is given for many types of dust by the equation

$$(dp/dt)_m = 0.7(p_m/p_i)^{2.5}, \qquad (3.3)$$

and the average rate by

$$(dp/dt)av = 0.3(p_m/p_i)^{2.5}. \qquad (3.4)$$

An investigation of maximum explosion pressures of dispersions of sugars in air was carried out by Meek and Dallavalle [3.10] in 1954, and has been reported by

Palmer [3.11] in a book on dust explosions and fires. The sugars were dextrose, sucrose and raffinose, and the rates of pressure rise (dp/dt) for any given pressure were found to be given by an equation of the form

$$(dp/dt) = Ap - Bp^2. \tag{3.5}$$

Experiments on peat dust explosions (peat can be thought of as geologically young coal) have been carried out in Finland, with varying particle size distributions and moisture content. At zero moisture content average particle diameters of 54, 96 and 165 microns gave rise to values of p_m of 8.4, 7.8 and 7.7 bar respectively, and values of $(dp/dt)_m$ of 610, 413 and 395 bar/sec. The times from ignition to peak pressure were 35, 47 and 45 msec. A moisture content of 14.1% by weight with an average particle diameter of 38 micron gave $p_m =$ 8.4 bar, $(dp/dt)_m = 513$ bar/sec. A moisture content of 33.6% by weight with a particle diameter of 72 micron gave values of 7.2 and 248 respectively. In reporting the Finnish experiments Kjäldman [3.12] also reported analytical work to model the explosive process. Essenhigh [3.13], however, has made the point that there has not been a great advancement in theoretical methods in recent times, and that most research since the late eighteenth century has been experimental. The main safety measure against dust explosions is the efficient venting of bins and containers, and this is dealt with in detail in reference [3.11] and in an industrial fellowship report issued by the Institution of Chemical Engineers [3.14].

In an appendix to his book, Palmer gives the explosion properties of a comprehensive selection of dusts, powders and particles, ranging from tea, tobacco, soap and rice to chemically complex cellulose, polyethylene and sodium powders. A selection of maximum explosion pressures in lb/in^2 and rates of pressure rise in lb/in^2/sec has been made by the author using Palmer's figures, and is given in Table 3.1.

Further information on pressures and pressure rise rates in relation to dust concentration has been summarized in ref. [2.38], using the earlier publication by Bartknecht [3.15].

3.3 GAS EXPLOSIONS

The best-known examples of gas explosions are associated with coal mining and the use of gas for domestic heating and lighting purposes. A major hazard in mining operations has always been the explosion of coal dust and firedamp (methane gas). Wherever coal is found underground there is likely to be methane, a product of the ageless decomposition of organic matter in coal, which is highly inflammable and difficult to detect. We are told that the 'working' of methane in a coal seam makes a sound like the faint humming of bees.

From the earliest time miners were conscious of the danger of using naked flames for light when working underground, and problems like this led to the well-known development of the safety lamp by Sir Humphrey Davy in 1815, and subsequently to the use of flickering oil lamps and carbide lamps. To help search for accumulations of methane gas, which is odourless and tasteless, the miners once took canaries underground, but more sophisticated gas detectors have been introduced over the years. These included a permissible flame safety lamp that would not become hot enough to ignite gas, but which would burn with a blue cap above the yellow flame if gas were present. The best answer to the gas hazard in mines is an efficient ventilation system, and most mines have a carefully planned system of passages, ducts, fans and air blowers to ensure that gases are diluted and driven out.

The accumulation of methane is not confined to coal workings, of course, and there have been examples of methane explosions in civil engineering workings

Table 3.1 Explosion pressures and rates of pressure rise of dusts, powders and particles

	Maximum explosion pressure lb/in²	Maximum rate of pressure rise lb/in²/sec
Asphalt	94	4800
Aspirin	87	7700
Bone meal	11	100
Charcoal	100	1800
Cinnamon	121	3900
Cocoa	69	1200
Coffee	38	150
Coffee (instant)	68	500
Corn cob	127	3700
Garlic	57	1300
Maize husk	75	700
Malt Barley	95	4400
Milk, skimmed	95	2300
Nylon	95	4000
Paper	96	3600
Pitch	88	6000
Polyethylene	80	7500
Rice	105	2700
Sawdust	97	2000
Soap	77	2800
Sugar	109	5000
Tea	93	1700
Tobacco, dried	85	1000
Walnut shell	121	5500
Wheat, flour	109	3700
Wood	90	5700
Yeast	123	3500

in very recent times. A notorious case was the explosion at the Abbeystead waterworks in the Lancashire fells in May 1984. An invisible cloud of methane gas rushed out of a tunnel that should have been full of water, a cigarette was lit and the subsequent explosion killed 16 people. Apparently the tunnel had been designed without ventilation shafts, and insufficient allowance was made for the possibility of pockets of methane finding their way into the tunnel. An important consideration was that under conditions of high pressure methane can be soluble in water. After the accident there was much scientific discussion about the source of the gas, and the Isotope Measurements Laboratory at Harwell found that most of the methane was over 20 000 years old, and had probably been lurking in rock fissures along the path of the tunnel under Lee Fell. The safety implications were profound, and resulted in legal action, discussion of which is beyond the scope of this book; the structural damage, which resulted in the destruction of the valve house, has been discussed in the journals of professional institutions.

Methane is a paraffin hydrocarbon, and is a product of the anaerobic bacterial decomposition of vegetable matter under water, hence the name marsh gas. The activated-sludge process of sewage disposal also produces methane-rich gases. It forms an explosive mixture with air when the methane (CH_4) content is between 5.3% and 13.9% by volume, and is the major constituent of commercial natural gas (after the extraction of gasoline), used for heat and power in industrial and domestic establishments.

This leads us to coal gas, which was the first domestic gas, produced when coal is heated out of contact with air to produce a stable residue, coke, and a gas with a heating value of about 500 BTUs per cubic foot. The coke residue was sometimes gasified in steam to produce 'water' gas. The first practical application for domestic use was developed by William Murdock at the end of the eighteenth century, in London. The famous Gas Light and Coke company was established in 1812. The main constituents of coal gas are hydrogen (about 50%), methane (about 25%), carbon monoxide (10%) and nitrogen (8%). These quantities varied a little depending on the precise design of the manufacturing retorts.

Coal gas was first used for cooking stoves by James Sharp of Northampton in 1830, and about thirty years later the gas stove industry had penetrated the USA. Gas for domestic lighting was overtaken by the electric light earlier this century, and the main domestic uses became cooking and heating. With this growth came the danger of domestic explosions. Eventually coal gas in the UK was replaced by the distribution of natural gas, and the local gasworks that were a feature of every country town have become obsolete in the past forty years or so. Natural gas (94% methane), produced from petroleum feedstocks, was used extensively in the USA, and began to be used in this country in the 1960s. It was relatively odourless, and the lack of the distinctive smell of coal gas meant that gas leaks were less likely to be detected. The odourization of natural gas from the North Sea is now undertaken at the storage installations on the mainland.

Explosions of domestic gas in the UK were running at about 100 per year in the 1970s and causing about 10 deaths per year. An inquiry into serious gas explosions, held in the late 1970s [3.16], noted that one-third of all explosions were caused by gas escaping from distribution mains and service pipes on the outside of buildings, and two-thirds from leaks at meters, appliances or installations within buildings. External gas escapes were frequently connected with the fracturing of cast-iron mains due to ground movement rather than corrosion. Often escaping gas in winter could not be released upwards because of frosted ground, and entered buildings through cellars or other service ducts. It was natural at the time for suggestions to arise that natural gas represented a greater explosive hazard than the old town gas, particularly as the countrywide distribution of natural gas meant an all-steel high pressure network. This comparison was investigated by an earlier inquiry [3.17], and it was concluded that the hazards from the use of both gas systems were similar. Natural gas has double the calorific value in BTU/ft^3, a much greater distribution pressure of about 30 mbar (about three times), and a lower maximum burning velocity (about half). Natural gas burner pressures are reduced by governing to about 21 mbar.

A domestic gas explosion is of the deflagration type, with a finite time from ignition to maximum energy release. The pressure rise is relatively small and direct injury due to this pressure increase (e.g. ear drum damage) is rare. Gas explosion fatalities usually occur from flying or falling objects, or by burns. Research has shown that from a total of 39 incidents, peak dynamic pressures from town gas explosions averaged 13 KN/m^2, and from natural gas explosions 11 KN/m^2. The air-rich mixture at which ignition occurs if a source of ignition is present is about the same for both gases at approximately 7.5% by volume of gas. There are important differences in flame propagation properties, however, with town gas having a propagation capacity about 5.5 times as great as for natural gas.

The effect of gas explosions in underwater tunnels is important, because the tunnel walls would usually be designed to resist external loads due to water and soil. In the event of an internal gas explosion the tunnel walls could experience a load reversal for which they had not been designed. The pressures generated by internal gas explosions have been investigated in Holland [3.18] and from these experiments the pressure-time history has been recorded. The duration of the overpressure plateau of 6 to 7 bar is governed by the time needed to vent the overpressure. The peak instantaneous pressure of 25 bar (2.5 N/mm^2 or 362 psi) represents a formidable internal load on the structure.

3.4 VAPOUR CLOUD EXPLOSIONS

Clouds of flammable vapour can be released accidentally from large process plants in the chemical and oil industries, particularly when the flammable materials are under pressure and above their atmospheric boiling points. If the leak of flammable material occurs for a sufficient time and rate for a significant amount to form an explosive mixture with air, then there is a strong possibility that an explosion will occur if there is a source of ignition. Experience shows that most major leaks occur in external piping systems, atmospheric vents, pumps and compressors, and that the most vulnerable part of any system is where smaller pipes and branches join large diameter, rigid pipes. If an explosion does not occur there is always the possibility of a flash fire or slow burning. Factors that influence the size and effect of a vapour cloud explosion include the amount of material in the cloud, the energy of the ignition source, turbulence, flame speed and wind direction. If a vapour cloud is above its upper flammability limit it can burn at a relatively slow rate; if it is below this limit but above the lower flammability limit an explosion can occur.

Because of the relatively slow pressure rise, and the absence of sudden shock, it is rather misleading to represent unconfined vapour cloud explosions by an equivalent amount of TNT. However, this is often done, and the evaluation of damage from explosions has been used to find equivalent values of peak instantaneous pressure and positive duration; it has been suggested that representative TNT explosions are $70 \, KN/m^2$ peak pressure, duration 20 ms, or $20 \, KN/m^2$ for a duration of 100 ms, depending on the span of the structural component under consideration. In a large unconfined explosion the pressure could rise to as much as 8 atmospheres over a period of one second.

There have been a number of catastrophic vapour cloud explosions in recent times, on land, offshore and at sea. Among the most well known occurred in 1974, when the cyclohexame plant at Flixborough, in the UK, exploded with devastating results and the death of 28 people. The leakage of flammable vapour came from damaged pipeworks. A second remarkable catastrophe was the explosion and fire at the Piper Alpha offshore oil platform in 1987, again due to the leakage of flammable vapour from pipework. The importance of accurate design and careful fabrication of structural pipes and their connections is clear. The inquiry into the Piper Alpha explosion noted the difficulties in making explosion hazard assessments of offshore platforms, because the explosion process is so difficult to analyse. Much of the available experimental data is derived from relatively small-scale tests, and it is not yet certain how far this can be applied to full-scale explosions. Further, most tests take place in vented chambers, or in chambers with obstacles, so the 'free field' conditions are not known.

The Piper Alpha technical investigation discussed the primary cause of the accident and considered in detail two likely explanations. It was clear, however, that there were two explosions, the first occurring in the gas compression equipment area. This caused a serious oil fire with large amounts of hydrocarbons to fuel it, which in turn caused the rupture of a major gas pipeline. This led to the initiation of the explosion and fireball that destroyed the installation. The preferred explanation for the first explosion was a gas leak from a condensate injection pump, caused by defective pipework.

The Flixborough explosion was intensely investigated, and it was found that a 20 inch diameter temporary connecting pipe, which had been installed with two misaligned expansion bellows, was the cause. About 45 tons of cyclohexame at a pressure of 125 lb/in^2 was released at a temperature of 155°C through two stub pipes. The release produced a large vapour cloud that was ignited at a furnace some distance from the source of the vapour, and there was a large fire before flame acceleration produced the explosion and blast wave. It is generally reckoned that predicted damage from an explosion of this nature can be estimated by assuming that the TNT equivalent for blast is about 2% of the combustion energy of the fuel, although at Flixborough the figure was as high as 5%. A very high value of 7.5% was calculated for an explosion in Missouri, USA, in 1970, when an underground pipeline containing propane burst, and the liquid formed a fountain above ground. The fuel-air mixture filled a large valley before exploding.

Another damaging explosion occurred in East St Louis, USA, in 1972, when a railway tanker wagon containing propylene hit an empty hopper wagon, the collision resulting in a puncture in the wall of the tanker. Liquid propylene was spilled along the tracks for 500 metres, the cloud ignited, and after an initial fire there was a very severe explosion that injured 176 people.

Another form of vapour explosion can occur when two liquids at different temperatures are mixed violently. The cooler of the liquids is vaporized very rapidly, and shock waves can be formed. Anderson and Armstrong [3.19] calculated relationships between the total heat released by cooling the hotter liquid to the initial temperature of the vaporizing cool liquid and the mass ratio of the cold liquid to the hot liquid. They showed that for a mixture of molten aluminium and water, the peak work in joules per gram of hot fluid, of about 700 units occurred at a mass ratio, gram of cold fluid to gram of hot fluid, of about 0.3. It is interesting to note that major natural disasters, such as the explosion of the island volcano of Krakatoa in 1883, are thought to be vapour explosions caused by the mixing of hot fluid (lava) and cold fluid (water).

Explosions can occur when fuel leaks into an enclosed space, mixes with air and forms a combustible mixture that is subsequently ignited. Another form of this type of explosion can occur when the space above the fuel in a container is in the explosive range, and is then ignited accidentally. Spectacular explosions in tanker ships have been recorded, and most of these took place in spaces that had a low length/diameter ratio. The rate of pressure rise is relatively slow, and the internal pressure will cause the plating to bulge outwards until tearing or fracture occurs. The blast wave produced is fairly weak. Explosions that are initiated in spaces with high length/diameter ratios can produce strong blast waves because flame propagation after ignition causes turbulence and a rapid pressure rise. Explosions in tankers and supertankers can be caused by ignition of escaping vapour during ballasting, but many tankers have been damaged by static spark ignition generated by the high pressure water spray used for cleaning. Well-documented explosions occurred in a Liberian tanker in Los

Angeles Harbour, USA, at Christmas 1976, and in the Norwegian tanker *King Haakon VII.*

3.5 THERMAL FLASH

A final characteristic of explosions must be examined before we end this brief survey.

Structural damage of a quite different nature can result from the extremely high temperatures created by the fireball of a nuclear explosion. The maximum fireball diameter for a 1 megaton bomb yield is about 1.4 miles, and for a 10 megaton weapon it would be 3.4 miles. The thermal flash from bombs of this magnitude can cause changes to properties of materials, and can weaken the resistance of metal structures that contain material of relatively low melting point. It might well be that an aluminium alloy structure would be robust enough to withstand the shock pressures and blast winds from a large nuclear explosion, but be seriously weakened because of a reduction in proof strength of the material.

For a low yield nuclear explosion, heat pulses at a range of 1000 ft from ground zero are of the order of 50 to 150 cals/cm^2. A heat pulse of 1000 cals/cm^2 is about equivalent to a maximum surface temperature in structural aluminium of 250°C, and for aluminium plating between 0 and 0.5 inches thick, this will lead to a reduction in strength by annealing. If a heat pulse of 100 cal/cm^2 is applied to the front face of a slab $\frac{1}{4}$ inch thick, the front and back faces will have both reached 250°C after 2 seconds. However, if the slab is $\frac{1}{2}$ inch thick, it has been shown by Parkes [3.20] that the behaviour is different. The front face, after reaching a maximum temperature of 200°C, falls gradually in temperature to reach a final value of 120°C after 2.5 seconds. Meanwhile the back face rises more gradually in temperature to reach the same final value.

By analysing structural members as a number of laminations, each with varying strain and properties, Parkes showed that it is possible to estimate the residual stress distribution after a period of uneven cooling. A high residual tensile stress means that a subsequent tensile loading of the member will cause failure at a relatively lower applied stress. Care should therefore be taken to inspect apparently undamaged aluminium alloy structures after a nuclear explosion to check for minor deflection and distortions that signify a noticeable reduction in structural strength.

3.6 REFERENCES

3.1   Fletcher, R. F. (1968) Characteristics of liquid propellant explosions, *Prevention and Protection against Accidental Explosion of Munitions, Fuels and Other Hazardous Mixtures*, New York Academy of Sciences, 432.
3.2   High, R. W. (1968) The saturn fireball, *Prevention and Protection against Accidental Explosion of Munitions, Fuels and Other Hazardous Mixtures*, New York Academy of Sciences, 441.

3.3 Chiotti, P. (1977) An overview of grain dust explosion problems, *Proc. Int. Symp. on Grain Dust Explosions*, Minneapolis, US Grain Elevator and Processing Society, 13.

3.4 Matkovic, M. (1977) Dust composition, concentration and its effects, *Proc. Int. Symp. on Grain Dust Explosions*, Minneapolis, US Grain Elevator and Processing Society, 62.

3.5 Tanaka, T. (1977) Predicting ignition temperature, minimum explosive limit and flame propagation velocity, *Proc. Int. Symp. on Grain Dust Explosions*, Minneapolis, US Grain Elevator and Processing Society, 79.

3.6 Parker, A. S. and Hottel, H. C. (1936) *Ind. Eng. Chem.*, **28**, 1334.

3.7 Cassel, H. M. and Liebman, I. (1959) *Combust Flame*, **3**, 467.

3.8 Kurosawa, M. (no date) *Chem A6*, **53**, 8584.

3.9 Enomoto, H. (1977) Explosion characteristics of agricultural dust clouds, *Proc. Int. Symp. on Grain Dust Explosions*, Minneapolis, US Grain Elevator and Processing Society.

3.10 Meek, R. L. and Dallavalle, J. M. (1954) *Ind. Eng. Chem.*, **46**, 763.

3.11 Palmer, K. N. (1973) *Dust Explosions and Fires*, Chapman & Hall, London.

3.12 Kjäldman, L. (1987) *Numerical Simulation of Peak Dust Explosions*, Technical Research Centre, Finland, Nuclear Engineering Laboratory, Research report 469.

3.13 Essenhigh, R. H. (1977) Problems of ignition and propagation of dust clouds, *Proc. Int. Symp. on Grain Dust Explosions*, Minneapolis, US Grain Elevator and Processing Society.

3.14 Schofield, C. (1984) *Guide to Dust Explosion Prevention, Part 1 – Venting*, Institution of Chemical Engineers, Rugby, England.

3.15 Bartknecht, W. (1982) *Explosions: Cause, Prevention and Protection*, Springer Verlag, Berlin.

3.16 *Report of the Inquiry into Serious Gas Explosions* (Chairman: P. J. King) (1977) HMSO, London.

3.17 *Report of the Inquiry into the Safety of Natural Gas as a Fuel* (Chairman: F. Morton) (1970) HMSO, London.

3.18 *Investigation of Transporting Dangerous Materials through Tunnels* (1982) Rijkswaterstaat, Directie Suizen en Stuwen, Utrecht, The Netherlands.

3.19 Anderson, R. P. and Armstrong, D. R. (no date) *Comparison between Vapor Explosion Models and Recent Experimental Results*, AIChE Symposium Series 138, Vol. 70, Heat Transfer Research and Design, 31.

3.20 Parkes, E. W. (1968) The inelastic behaviour of aluminium alloy tension members when subjected to heating on one face. In Heyman and Leckie (eds), *Engineering Plasticity*, Cambridge University Press.

4

Structural loading from distant explosions

The dynamic loading of structures from detonating explosions is due to the instantaneous or very fast increase in air pressure associated with the shock front, and to the transient forces associated with the blast winds that follow the passage of the shock front. So far we have limited our survey mainly to the 'free field' conditions that result from many types of explosion, but now we must introduce into this clear space structures of varying size, shape, flexibility and ductility. Some of these structures will be sufficiently stiff, or be supported in such a way that the explosive actions will not be influenced by structural behaviour under load. Other structures will deform or move in such a way that their behaviour influences the loading they receive. In the jargon of the analyst the former are decoupled, the latter coupled. Before we can consider this further we must review the pressures that arise when shock fronts are reflected or refracted by obstacles, above and below ground and under the sea.

There are three settings to be considered. In the first the explosion is large and distant, so that the loading actions can be considered uniform over the faces of the structure. The Mach stem from a nuclear explosion, for example, travelling over the surface of the ground, would load the component faces of a structure in its path with a reasonably uniform distribution. The second setting, considered in Chapter 5, is when the explosion is relatively small and occurs close to, but not in contact with, a structural member. The major effects of the explosion are concentrated over a relatively small area of the surface, and the loading actions decrease in magnitude rapidly outside this area. Impulsive loads vary with distance along or across the member, and the time at which they begin is also gradually increasing with distance. These we will term concentrated loading actions; in some analytical schools these actions are designated as being 'local-ized', an ungainly adjective. The third category is the contact explosion, considered later, when the exploding charge is either in direct contact with the structure or has been buried in the material of the structure by static or dynamic penetration. An example of static penetration is the boring of holes into rock and the careful placing of a charge at the end of the hole. Dynamic penetration occurs when the warhead of a bomb or missile is contained in a shaped container that hits a structure at high velocity and enters the component before exploding. The component can have the solidity of a reinforced concrete slab, the cellular construction of an aircraft body, or the laminations of armour plate.

The loading of structures by an extremely concentrated action that damages the fabric by a cutting action must also be considered. This leads us later to the consideration of hollowed or shaped charges that project molten metal linings into the structure like jets. This chapter, however, is confined to one distant explosion, and to the analytical history of the investigation of structural loading when the loading actions are assumed to be uniform. Much of this history has been set down and explained with diagrams in many earlier publications, and there is no point in repeating the details here, so what follows is a review of the major elements of the loading analysis.

4.1 UNIFORMLY DISTRIBUTED PRESSURES

The fundamentals of reflected shock fronts were discussed briefly in section 2.1, but it is now necessary to look more closely at reflections from above-ground structures and components. An example often considered is the action of a shock front from a detonation on a surface structure of simple rectangular box-shape, having one face normal to the direction of propagation of a shock wave from a distant explosion. If the explosion is directly above the structure, the shock wave will strike the upper face and a reflected shock is formed. The overpressure on this face rises to p_r, which as we saw in Eq. (4.1) is given by

$$p_r/p_0 = 2(7p_a + 4p_0)/(7p_a + p_0), \tag{4.1}$$

where p_0 is the peak 'free field' incident pressure, p_a is atmospheric pressure, and the angle of incidence of the shock front is zero. After the shock wave has impinged on the surface of the earth adjacent to the structure, the sides of the structure will be loaded by the same reflected pressure (which acts in all directions). It is assumed in this analysis that the wavelength considerably exceeds the dimensions of the structure.

If the explosion is at ground level, and the centre of the explosion lies on the normal to the side face of the structure at its mid-point, then the shock wave will strike the side face normally and a reflected shock is formed having an instantaneous pressure p_r. At the moment this reflected front is formed, the initial non-reflected pressure p_0, just above the top edge of the front face, initiates a rarefaction wave, travelling at the speed of sound, which heads down the front face towards the ground. This causes disintegration of the reflected shock, until it is in equilibrium with the dynamic pressures due to the following blast winds, q.

Meanwhile, at a time equal to the length of the structure divided by the front velocity, $\bar{u}$, the incident shock front travels over the top and along the sides of the box and reaches the rear edge. It then spills down the back wall. On the flat roof and sides the instantaneous pressure rise is to p_0 (not p_r), but because of the pressure difference at the junctions of the top and sides with the front face, vortices will be formed there. See, for example, the book by Norris *et al.* [4.1].

In determining the loads on the structure it is useful to consider the instant at which the shock front strikes the front face as $t = 0$, and the time for reflection effects to clear the front face as $t = t_c$, where

$$t_c = 3s/\bar{u}. \tag{4.2}$$

In this equation s is the clearing height, taken as either the half width of the front face $(B/2)$ or the full height (H), whichever is the lesser, and $\bar{u}$ is the velocity of the shock front. Once the time t_c is reached, the pressure on the front face (taken as uniform) is a combination of the gradually decaying incident pressure, p, and the dynamic pressure, q. The variation of dynamic pressure with time is usually given by the equation

$$q/q_0 = (1 - t/t_0)\, e^{-3.5t/t_0}, \tag{4.3}$$

which is similar to the expression for overpressure decay but with a different decay rate. From Eq. (1.28) we note once more that the dynamic pressure in psi at time t_0 is given by

$$q_0 = 14.7\left[\frac{(5/14)(p_0/14.7)^2}{1 + (1/7)(p_0/14.7)}\right]. \tag{4.4}$$

To obtain a true measure of the dynamic loading due to the blast winds on the structure, it is necessary to multiply q by the drag coefficient, C_d, for the shape under consideration. So the total front wall pressure (p_s) from time t_c onwards is given by

$$p_s = p + C_d q. \tag{4.5}$$

For the front face of a rectangular box structure C_d is often taken as 1.0, although wind tunnel tests show that the average pressure due to blast winds is rather less than 1.0 q. Some analysts therefore suggest $C_d = 0.85$ for the front face.

When the shock wave reaches the rear wall of the structure distant L from the front wall, vorticity again occurs at the edges, causing a reduction in incident pressure due to suction. According to some authorities the overpressure on the rear wall reaches a maximum in a time given approximately by

$$t_L = (L + 4S)/\bar{u}. \tag{4.6}$$

The sides and top of our simplified box are not fully loaded until the shock front has travelled the full length of the structure. The average pressure is therefore considered to be the shock overpressure plus the drag loading at a distance $L/2$ from the front face. The loading increases from zero to the average value in a time $L/2\bar{u}$, so that, for the top and side faces

$$p_s = [p + C_d q]_{t=L/2\bar{u}}, \tag{4.7}$$

and for the rear face

$$p_s = [p + C_d q]_{t=(L+45)/\bar{u}}. \tag{4.8}$$

Values of C_d for the side, top and rear faces are known to vary with the dynamic pressure, q. In 1961 Newmark *et al.* [4.2] suggested that C_d should be -0.4 for $0 < q < 25$ psi, -0.3 for $25 < q < 50$ psi, and -0.2 for $50 < q < 130$ psi. Note that in all these circumstances c_d is negative so that $(p + C_d q)$ is less than p.

It is sometimes useful to compute the horizontal loading on a structure, and this is best done graphically to take account of the phase displacement of the arrival of the peak overpressures. The pressure/time diagram is constructed from the information in Eq. [4.5] and [4.8], and this leads to an indication of the net horizontal loading at any time t.

If the shock front, whether from above or from the side, does not engulf the structure normal to the top or front faces, then it is necessary to resolve the pressures in directions normal to the walls of the structure, carry out calculations for each direction, and add the results. This procedure still assumes that the explosion is large enough, and far enough away to produce uniform loading conditions. Values of p_r for various angles of incidence have already been given in Figure 2.8.

The enveloping of the structure by the shock wave is a diffraction process, and for large structures under short duration loads this process is more important than blast wind loading. When the structures are relatively small and the dynamic pressure duration long, drag loading becomes more important. Structures are therefore sometimes referred to as diffraction or drag targets, depending on the above factors.

4.2 LOADING/TIME RELATIONSHIPS

Some simple above-ground protective structures are, in fact, rectangular boxes, but this is exceptional. Complex structural shapes such as aircraft, ships or vehicles, or domed and cylindrical bunkers are mainly curved in plan and elevation. The computation of shock and blast loading then becomes a difficult and tedious process. The reflection varies from point to point over the surface, and the time of decay to stagnation pressure depends on the location of the point. The drag coefficient varies with structural geometry, and is sometimes given as an average coefficient for the whole complex shape; at other times it is given as a local coefficient for component details. Certain structures suffer local damage from the shock wave (e.g. blown-out windows), which alters their behaviour under dynamic wind loads; others have openings that allow the blast wave to enter the inside of the structure and alter the net loading on the faces.

Table 4.1

Member	C_d
Circular cylinder, side-on	1.2
Sphere	0.47
Cylinder, end-on	0.82
Disc, face-on	1.17
Cube, face-on	1.05
Cube, edge-on	0.80
Long plank shape, thin face-on	2.05
Long plank shape, edge-on	1.55
Narrow strip, face-on	1.98

Drag coefficients for bodies in the subsonic flow range have been given by Hoerner [4.3] who lists the values shown in Table 4.1 for C_d for some simple shapes (see also Baker *et al.* [4.4]).

Values of C_d for many other shapes, including framed structures such as masts and towers are given in books on the effect of wind on structures.

Glasstone and Dolan [4.5] gave procedures for deriving air blast loading as a function of time for partially open box structures, framed structures and cylindrical and arched structures, in addition to the simple box structure considered above. They define a partially open rectangular box structure as one in which the front and back walls have about 30% by area of openings or windows, and have no interior partitions that might influence blast wave behaviour. The average loading on the front face is influenced by the fact that the value of t_c is now $3s^1/\bar{u}$, where s^1 is the average distance from the centre of a front face wall section to an edge of wall at an opening. The internal pressure begins to rise at zero time because the blast wave enters the openings, eventually reaching the blast wave overpressure. The dynamic pressures are assumed to be negligible (i.e $q = 0$) on the interior. The average loading on the sides and top are similar to those for a closed box, but the inside pressures take a time of $2L/\bar{u}$ to reach the blast wave value. For much of the decay period the internal pressure is greater than the external pressure.

The outside pressures on the back face are similar to those for a closed structure, except that s is replaced by s^1 in the analysis. The inside pressure reaches a similar value to that of the blast pressure, instantaneously, at a time $L/\bar{u}$, and then decays. The dynamic pressure is reckoned to be negligible.

When a blast wave meets a cylindrical structure there is a very complex interaction with the curved surface, but the loading increases from zero to a maximum when the blast front has traversed one radius; this occurs at a time $D/2\bar{u}$ (where D is the cylinder diameter). It is sometimes assumed that the

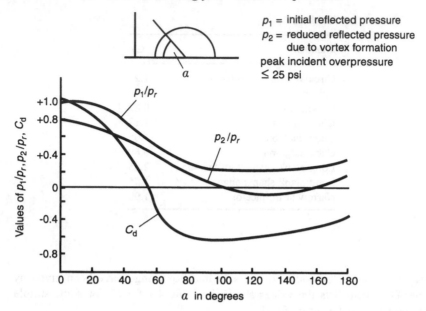

Figure 4.1 Blast wave meeting 180° semi-circular arch (from Newmark *et al.*, ref. 4.2).

maximum average pressure on the front of the cylinder is $2p$, and that this decays linearly until a time $2D/\bar{u}$. The average side loading is given as

$$\frac{p}{2}\left(\frac{3D}{2\bar{u}}\right)$$

in ref. [4.5], and the maximum back surface loading as

$$p(20D/\bar{u}) + C_d q(20D/\bar{u}).$$

When a semi-cylindrical, or arched structure is loaded by a shock wave perpendicular to its longitudinal axis, vortex formations can occur immediately after reflection, so that a temporary sharp drop can occur before the stagnation pressure is reached. Subsequently the total pressure, $p + C_d q$, decays in the usual way. The relationship between p_1, p_2, C_d and the angle α for a semi-circular arch is shown in Figure 4.1, where p_1 and p_2 are related to p_r, the ideal reflected pressure for a free surface. Since the structural loads are normal to the surface, the total horizontal load on the structure is formed by summing all horizontal components, and Figure 4.1 gives the approximate equivalent net horizontal force.

Many arched structures are not fully semi-cylindrical, and for 120° arches similar relationships to those in Figure 4.1 have been presented (see, for example, ref. [4.2]). It can be seen from these relationships that C_d varies from positive to negative as the angle increases, and is also influenced by the peak incident overpressure. The variations in C_d are partly influenced by the proximity of the ground surface to the boundary layer. Blast loading on arches was investigated by F. E. Anderson of the US Defense Atomic Support Agency in 1958, and his work was used by Newmark and others in drawing up the figures.

Anderson also investigated domes, and the results for 45° domes are given in Figure 4.2 when the peak incident overpressure, p_0, is less than 25 psi. The variation of C_d along the line of symmetry of the dome shows that $C_d = 0$ when the angle α is about 70°. Along a transverse line of a 60° cylindrical arch under similar loading functions C_d is zero when α is about 55°.

In the discussion on the loading of a semi-circular cylinder it was assumed that the shock front was travelling in a direction normal to the longitudinal axis of the cylinder. If, however, the shock travels in the same direction as this axis and first impinges on the flat end of the structure, the values of pressure ratio and drag coefficient will be similar to those for a rectangular box structure.

All the above loading analyses resulted from research in connection with the design of structures to resist nuclear weapons effects, so it was presented, with

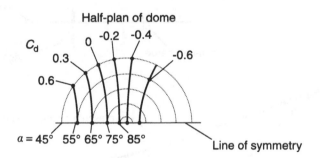

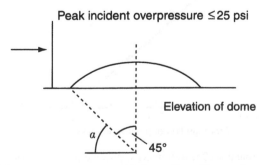

Figure 4.2 Blast wave meeting 45° dome (from Newmark *et al.*, ref. 4.2).

Structural loading from distant explosions

revisions, during the period 1950 to 1980. However, earlier assessments of uniform structural loading were included in the work of Christopherson [4.6] in 1945. He investigated the loads on a 10 ft high wall situated normal to the direction of travel of a blast wave from a 4000 lb bomb, in an illustration of the numerical application of diffraction theory. The duration of positive impulse was 55 in sec, giving a wavelength of 63 feet. The total depth of the wall and its mirror image when reflected in the ground was 20 feet, which was about 0.32 times the wavelength. The total pressure on the back of the wall was computed and it was found that the maximum pressure in the diffracted wave was about 0.6 times the incident pressure. By integrating over the front and back surfaces the mean pressure time curve of the loading action tending to overturn the wall was found.

Figure 4.3 shows this pressure–time relationship together with the pressure–time curve for a plane of infinite extent in all directions. This comparison established that the duration of the forces is reduced to one-quarter of the positive duration of the blast by diffraction, and that this reduced duration is less

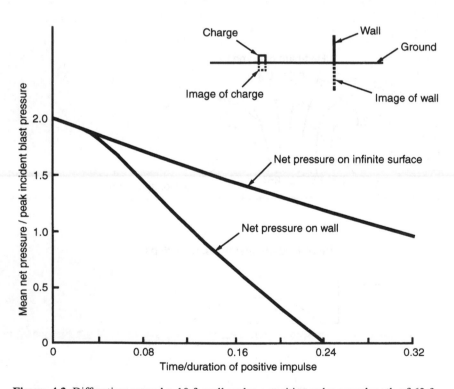

Figure 4.3 Diffraction around a 10 ft wall under a positive pulse wavelength of 63 ft (from Chrisopherson, ref. 4.6).

than the time for a sound wave to traverse the face of the wall from side to side. This example illustrates two general principles spelt out by Christopherson:

(a) An initial peak pressure is not diffracted without loss, so an obstacle casts a distinct shadow when loaded by high pressures of short duration.
(b) The duration of impulses from a shock wave of small amplitude, acting on a body having dimensions that are small compared with wavelength, will be governed by the time taken for the sound wave to travel around the obstacle, rather than by the duration of the original pulse.

This means that weak shock waves will have relatively little effect on columns, chimneys or the members of frameworks that can be quickly enveloped by the diffracted shock front. When the shock wave has a very large amplitude the first principle (a) is still true, but the second will not apply.

The blastwave loading on open structures such as a bridge truss or a lattice tower is caused almost entirely by the drag pressures due to the high-velocity blast winds, and it is usual to neglect the overpressure loading due to the shock front because of the very short duration of its effect on individual truss members. As before, the drag pressure, p_a, is given by $p_d = C_d q$. A frame structure made from round tubes has a lower drag coefficient on individual members than a structure made from structural sections that consist of assemblies of flat webs, flanges or plates. In general the latter have values of C_d in the range 1.8 to 2.2, whereas the round tubes have values of C_d in the range 1.0 to 1.2.

When structural members are in close proximity, shielding has an effect on loading, as illustrated by the shielding of the leeward truss by the windward truss in a bridge. Figure 4.4 gives the ratio of leeward to windward drag coefficients in terms of 'solidity' ratio. This is the ratio between the total area of individual members in cross section to the area of a completely solid cross section having the same overall dimension, and it assumes that the members are set normal to the blast wind direction.

The effect of shielding on the blast loading of circular cylinders was investigated by Mellsen [4.7] in 1973. Two cylindrical cantilever beams, each 1.5 inches diameter, were subjected to a blast wave of 20 psi incident overpressure and 25 msec positive duration. The blast wave travelled in a direction so that one cylinder exactly shielded the other. The distance between the cylinders was increased from 2 to 8 diameters in a series of tests, and ratios of peak strains were measured. The ratio of rear cylinder peak strain to front cylinder peak strain increased from 0.4 to 0.8 over a range of 2 to 8 diameters, but for a greater distance apart of 16 diameters there was only a small further increase in peak strain ratio to about 0.9. The tests also indicated that the loading ratio depended on the value of peak incident blast pressure. When this pressure was decreased incrementally from 20 psi to 13.5 psi the ratio rose linearly from 0.8 to 0.95 for a cylinder spacing of 8 diameters.

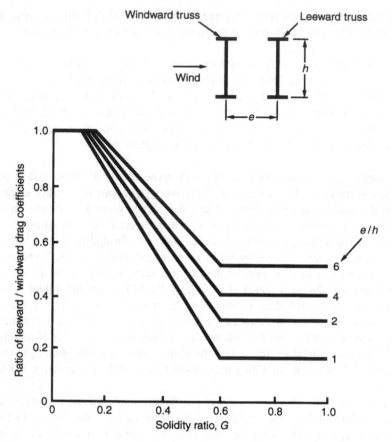

Figure 4.4 Relationship between drag coefficient ratio and solidity ratio for two parallel trusses.

4.3 LOADS ON UNDERGROUND STRUCTURES

There are several ways in which a structure buried in soil can be loaded by large and relatively distant detonated explosions. The structure may be buried at a sufficient depth beneath the surface to escape the effects of air blast dynamic pressures from large above-ground explosions. Structural loading can then result from the surface pressures due to the instantaneous pressure rise in the shock front, causing direct ground shock, or from air-induced ground shock which arises from the passage of the shock front across the surface of the soil above the structure. If the large explosion takes place on the surface of the soil (a contact surface burst) the earth shock can travel directly through the soil on its way to the structure. Direct earth shock can also emanate from an explosion that occurs well below the surface of the soil.

Shallow or semi-buried structures, although covered with soil, can be partly above ground level, and because of this can also be loaded by the blast winds. If the soil surrounds the structure and rests at its natural angle of repose, its presence will considerably reduce the value of the coefficient of drag on the front face and at the same time buttress the structure against lateral movement.

Underground structural loading is also influenced by the flexibility of the structure in relation to the properties of the surrounding soil and the soil-structure interaction effects that occur as the structure begins to deflect under shock pressures. The structure and its surrounding soil form a composite body, the loading on which can be influenced by soil-arching and the disturbance to the natural properties of the soil.

Much of the early work in this field was driven by the need to assess nuclear weapon effects, and there are good summaries of the problem in the reports by Newmark and others, and by Glasstone and Dolan, discussed earlier in references [4.2] and [4.5]. Using their work as a basis, we will discuss first the structural loading that comes from blast-induced ground shock. We saw earlier that there are two types of air-blast-induced ground motion. In that where the velocity v of the shock front exceeds the dilational seismic velocity (u) of the soil, there is a sloping shock front below ground. If $v = u$, the ground shock front becomes almost vertical near the surface, and if $v = u$, the shock front in the ground outruns that in the air. The relationship of pressure with duration will attenuate as greater depths of soil are considered, and Figure 4.5, from the work of Newmark [4.8] illustrates this. As the depth increases the rise time is longer, there is a lower peak pressure and a longer decay. The total impulse (i.e. area under the pressure–time curve) remains about the same, and it was established experimentally that the rise time at any depth was about half the time of transit of the shock front to the point considered. The changes in shape of the pressure pulse were a direct result of the stress-strain characteristics of the soil in unidirectional compression. A simplified empirical equation was proposed by

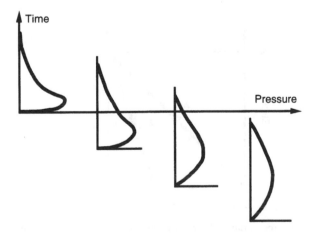

Figure 4.5 Attenuation of pressure wave with depth (from Newmark, ref. 4.8).

Newmark and Haltiwanger [4.9], which gave reasonable agreement with test results obtained by Newmark and Hall [4.10]. The relationship between peak vertical stress, σ_z, and depth z, and the peak surface overpressure on the surface, p_s, was given by

$$\sigma_z = \alpha p_s, \tag{4.9}$$

and α was the peak stress attenuation factor, defined as

$$\alpha = (1 + z/L_W)^{-1}, \tag{4.10}$$

where

$$L_W = 230 \text{ ft} \left(\frac{100}{p_s}\right)^{1/2} \left(\frac{W}{1MT}\right)^{1/3}. \tag{4.11}$$

W = yield of explosion in megatons of TNT (MT), and p_s was expressed in psi. Further analytical work in the 1960s by Hendron and Auld [4.11] suggested that the above equation was only applicable for a soil with a strain recovery ratio of $\frac{1}{3}$, and a propagation velocity of peak stress for initial loading of 622 ft/sec. If the unidirectional stress-strain character of a soil can be generalized by the bilinear hysteresis-type curve shown in Figure 4.6, where M_L is the loading modulus and M_U is the unloading modulus, then the propagation velocity, V_L, is given by

$$V_L = (M_L/\rho)^{1/2}, \tag{4.12}$$

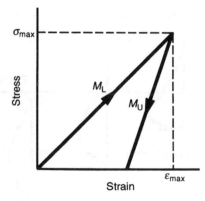

Figure 4.6 Bilinear stress-strain relationship for soil (from Hendron and Auld, ref. 4.11).

where ρ is the mass density of the soil. The strain recovery ratio, M_L/M_U, could well lie in the range 0.3 to 0.7 and to take account of this greater range of conditions Hendron and Auld proposed an amended equation

$$L_W = 230 \text{ ft} \left(\frac{100}{p_s}\right)^{1/2} \left(\frac{W}{1MT}\right)^{1/3} \left(\frac{V_L}{622}\right), \tag{4.13}$$

where V_L was in ft/sec. For a given peak overpressure and strain recovery ratio α scaled as $z/w^{1/3} \cdot V_L$. A further discussion of wave propagation in soils and a review of elementary wave theory was given at about the same time by Allgood [4.12].

It is generally accepted that when structures are buried very deeply and subjected only to direct ground shock, the loading action due to this shock is insignificant when compared with the loads from the dead weight of soil cover. However, if a huge megaton range nuclear bomb were to burst directly over a deep structure, it could cause extensive damage at depths of 500 feet and over.

The ground shock wave from a very deep underground nuclear explosion gradually weakens into a train of seismic waves that can cause ground motions and earthquake-type damage a long way from the point of detonation. Structural loading is influenced by the type of soil in which the structure is buried, and by the soil-structure interaction effects associated with lateral ground motion. As with all analytical work on buried structures, care must be taken when using measurements taken under static loads that they are still applicable in dynamic and vibrational conditions.

Design guides for underground structures under blast and shock loads, see for example ref. [4.2], often distinguish between shallow-buried, surface flush and earth-mounded structures. Surface flush structures have soil covers (h) over the highest point of their upper surfaces in the range $0 < h < 0.2l$, where l is the clear span between supports of the roof slab, crown arch, dome or shell as the case may be. Shallow-buried structures have covers in the range $0.2l < h < 1.5l$, which is reckoned to be deep enough for dynamic soil arching to influence the loads on the structure. Unfortunately, although there is considerable experimental and analytical knowledge on soil arching under static conditions, much less is known about dynamic arching. Knowledge on static soil arching has been discussed by the author in an earlier book [4.13], and it has been shown that for most soils shear strength is the key factor. The high shear strength of a well-compacted granular soil leads to considerable static arching, whereas the low shear strength of soft clays results in negligible arching. Under dynamic conditions, experiments by the author [4.14] suggested that static shear strength was no longer a factor, and that granular soils behaved very similarly to soft clay soils. The static redistribution of loads away from flexible members to their stiff supports can considerably reduce load effects and lead to increased structural resistance. Under dynamic conditions, however, shear stresses cannot develop quickly, and soil arching cannot redistribute loads· until a rarefaction wave travels from the rigid to the flexible areas of the structure. Because of the general uncertainty it has been thought conservative to ignore soil arching when assessing the dynamic loads on shallow-buried structures from large and distant explosions. This can lead to uneconomic designs, but research to investigate dynamic arching more thoroughly has been limited by a reduction in funding due to a lesser nuclear threat in recent years.

The loading of shallow-buried structures is also influenced by the reflection of the incident shock wave under soil-structure conditions. We saw earlier that the surfaces of above-ground structures cause large reflection pressures, p_r, that can be double the incident pressure, p_0. In underground situations the reflected stress, σ_r, can be influenced by movement of the structure in relation to the soil, by the rarefaction wave travelling from the rigid to the flexible areas of the structure, and by the possible reflection of the same waves downwards by the soil/air surface. These factors affect the time after the initial shock meets the structure that the pressure–duration curve returns to incident pressure decay values (i.e. a time t_d corresponding to time t_c in the above ground structural analysis). For the particular case of a reinforced concrete flat slab roof, having a thickness and a seismic velocity in the concrete of v_c, it has been suggested that $t_d = 12T/v_c$, because the tensile wave reflecting back from the underside of the roof will influence the conditions of reflectivity at the upperside.

Surface-flush underground structures ($h < 0.2l$) have a slightly different loading condition, because the duration, t_d, of reflected stress is smaller. This is because a relief shock wave can quickly be reflected downwards from the ground surface, and t_d now becomes the smaller of the values h/v_s or $12T/v_c$ for flat concrete slabs, where v_s is the seismic velocity in the very thin soil cover.

Shallow-buried and surface-flush structures are normally buried deeply enough to avoid dynamic wind loads, but this is not the case for mounded structures. The drag coefficients for mounded structures with 1 in 4 slopes on the berms are about 1.0 for the slope facing the direction of the blast, and −0.2 on inclines facing directly away from the blast, so significant drag and reflection can occur for this gentle slope. Nevertheless, in the initial stages of structural design a 1 in 4 slope is often considered sufficiently shallow to ensure that surface blast winds do not load the structure, and the loading is restricted to the attenuation of the instantaneous overpressure p_0 by the small thickness of soil above the roof or crown of the structure.

In all underground structures the position of the water table is important. For shallow-buried surface-flush or mounded structures it is not advisable to set the structure under the water table, as the saturated soil will result in higher loads. In assessing structural loading the possibility of a high water table coinciding with the detonation of a large explosion should be the subject of a risk analysis. In mounded structures it is likely that the soil conditions will change within the depth of the structure due to the change from backfill soil to undisturbed soil to waterlogged soil. Underground structures often have surface structures above them, and in certain circumstances the presence of such structures can affect the peak incident overpressure reaching the underground structure for a large air burst.

It is clear that the loading on underground structures is influenced by the shape and flexibility of the structure. For example, because of the support that can be mobilized from well-compacted soil, it is possible to use thin-walled metal construction for underground structures or for the linings of temporary trench shelters or earthworks. Here, the structure is formed from thin sheet elements that in themselves have low structural strength. but which can support large overpressures if properly embedded in soil. Corrugated steel culvert material used in civil engineering is a good example. The limitation to load-carrying ability is often the inward buckling of the walls of the structure, or the tensile rupturing of sheeting due to large membrane forces, rather than collapse by the formation of ultimate moments in structural members.

For heavier, thick-walled construction, typical of reinforced concrete, there are broad, approximate quasi-static loading conditions that can be used for initial design calculations without recourse to complex analysis or personal computers. For example, in a shallow-buried rectangular structure, with a cover depth less than one-half the span of the roof slab, it can be assumed roughly that the roof and floors are loaded with the ground surface peak overpressure, p_0. In a dry soil the effect of attenuation might limit the side pressure on the walls to $0.5 p_0$, but in saturated conditions with a high water table this would increase to p_0. The loading conditions on roof elements would be the full overpressure, p_0, plus the dead load of the roof structure, the soil cover, and any debris that might fall on the surface. For side wall elements the static earth and water pressure would need to be added to the blast overpressure. Examples of reinforced concrete underground shelters to withstand unclear explosions, designed on this rough quasi-static basis, were given in the UK Home Office publication, *Domestic Nuclear Shelters* [4.15]. It is interesting to note that in the UK domestic nuclear shelter design of the 1960s and 1970s the loading overpressures were limited to three atmospheres (315 kp_a or 45 psi), because at pressures above this the effects of Initial Nuclear Radiation would probably kill the occupants.

Perhaps we should pause for a moment and consider radiation doses, since they are a fundamental part of nuclear explosions. They are measured in various units such as roentgen, gen, rad or rem, depending on the precise kind of radiation. We will only consider the roentgen here. Acute radiation sickness is produced by a brief exposure of the whole body to 50 roentgens; for exposures between 50 and 200 there could be weakness and fatigue in addition to sickness, but only about one person in twenty would need medical attention. Between 200 and 450 roentgens there would be moderately severe illness, and perhaps one quarter of the people so exposed could die. Exposures of over 600 roentgens would lead to death in less than 14 days in almost every case. Exposures in excess of several thousands causes severe brain damage and death within hours.

Table 4.2

Overpressure (psi)	45	15	11	6
Radiation for 1 MT (roentgens)	70 000	10 000	250	2
Radiation for 20 MT (roentgens)	9000	1	–	–

The intensity of Initial Nuclear Radiation falls off rapidly with distance from the centre of the explosion, and because of this the relationship between blast shock pressure and radiation is governed by bomb size. For low-yield explosions the lethal range of radiation extends beyond the range of lethal blast, and it is possible for a person in the open to survive the shock and blast winds but be killed by radiation. For very large megaton explosions people can be killed by blast winds when only receiving a negligible radiation dose. This is illustrated by Table 4.2.

As a general rule, underground structures designed to protect the occupants against peak overpressures of 11 psi (77 kp_a) and above from a nuclear bomb should also be checked against radiation. In most cases the thickness of cover over the structure to give sufficient protection against the blast and shock is also sufficient to reduce radiation to acceptable levels for the occupants.

4.4 LOADS ON UNDERWATER STRUCTURES

There was a fair amount of openly reported research on underwater loading during the latter stages of the Second World War, when the behaviour of naval ships and dock installations under nuclear shock was investigated, but since then much of the work has been shrouded in confidentiality. Reference [4.16], published by the US government in 1946, addressed the problem of underwater shock in relation to naval structures and gave information on underwater reflectivity. It was concluded that the principles of reflectivity are similar to those for above-ground structures, so that an underwater incident wave imping-ing on a flat surface at an angle of incidence of zero would be reflected with a pressure increase of almost double. If the shock wave strikes the structure at an angle it is possible to predict theoretically a lower limit to the incident pressure above which acoustic-type reflection cannot occur. As with above-ground struc-tures, it is possible for waves striking the structure at an angle to produce higher reflected pressures than at zero incidence.

For rigid underwater structures the structural loading is not influenced by the flexibility of the structure, but when a shock wave meets the relatively flexible surface of ships' plating the applied pressure can be reduced by the presence of a rarefaction wave initiated by the deflection of the curved plates. Should the reduction be so great that the applied pressure drops below zero, cavitation can occur. If the shock wave has a relatively long duration, and meets a rigid structure with a short natural period, then the loading would be similar to that from a steady, static pressure, and the maximum reflected pressure would be the basis of damage assessment. If the wave has a short duration compared with the natural period of the structure, which is more likely when the structure is very flexible, then the basis of damage assessment would be the total impulse of the shock load.

We have already seen that the pulsating bubble phenomena associated with underwater explosions result in the mass movement of water as well as pressure pulses. The pressure pulse from a bubble at its minimum is greatest when bubble migration is small. According to ref. [4.16] this meant for example that for a 300 lb charge of TNT in a mine, the mine should be moored 14 feet above the seabed. The mass motion of water that accelerates radially outwards from the expanding bubble is an important factor when considering close-range underwater damage, but for larger, distant explosions its effect is considerably diminished. This is because the kinetic energy of the outflowing water falls away as the fourth power of distance, whereas the energy in the shock wave is reduced according to the second power of distance.

Christopherson [4.6], in summarizing the work in underwater explosions in the UK during the Second World War, made the point that underwater structures such as submarines are virtually unsupported, in that there is no rigid support to ensure that the natural period of oscillation of the vessel is short in comparison with the duration time of a high incident pressure. He quoted the work of G. I. Taylor [4.17] who considered the oblique impact of an underwater shock wave on targets with a variety of supporting forces. For the simplest condition, in which a shock wave strikes normally a completely unsupported underwater plane surface having mass m per unit area, he showed that on the surface of the plate the total pressure (p) is given by

$$p/p_1 = e^{-t/\theta} + \phi(t),$$
(4.14)

where p_1 is the pressure in the incident shock wave, t is the time measured from the moment of arrival, and θ is the time for the pressure to fall to half its initial value. If the unsupported plane surface remains in contact with the water in the incident wave it is possible to formulate its equation of motion, and from this to show that

$$p/p_1 = \frac{2}{\varepsilon - 1} [\varepsilon e^{-\varepsilon t/\theta} - e^{-t/\theta}],$$
(4.15)

where ε is the dimensionless parameter $pc\theta/m$, c is the velocity of sound in water, and ρ is water density. The maximum velocity of the plane surface occurs when the pressure has fallen to zero, i.e. when

$$t/\theta = \frac{1}{\varepsilon - 1} \cdot \log_e \varepsilon,$$
(4.16)

and at this moment the water begins to exert a retarding force on the plane surface. The plane surface then loses contact with the water and cavitation occurs. Substituting for t/θ in [4.15], and remembering that the net velocity (V)

at the target plane is found from the equation $\rho c V / p_1 = e^{-t/\theta} - \phi(t)$, leads to the expression

$$V = \frac{2p_1\theta}{m}\, \varepsilon^{1/(1-\varepsilon)}. \tag{4.17}$$

The energy imparted to the target plane is $\frac{1}{2}mV^2$, which can be written as

$$\tfrac{1}{2}mV^2 = (2p_1^2\theta^2/m) \cdot e^{2/(1-\varepsilon)}, \tag{4.18}$$

and if this is available to load the target then the extent of damage can be predicted.

Christopherson suggested that underwater loading (and hence damage) will result from the following sources:

(a) The initial shock wave, acting for a very small time of perhaps 1 msec.
(b) The following kinetic wave, which becomes important after the initial shock wave ceases to act, but only lasts about $0.02W^{1/3}$ sec (W in lb).
(c) The second and later bubble expansions, which are only important when the bubble approaches the plane surface between expansions.

4.5 REFERENCES

4.1 Norris, C. H. *et al.* (1959) *Structural Design for Dynamic Loads*, McGraw Hill, New York.
4.2 Newmark, N. M. *et al.* (1961) Design of structures to resist nuclear weapon effects, *Manual of Engineering Practice*, No. 42, American Society of Civil Engineers.
4.3 Hoerner, S. F. (1959) *Fluid-Dynamic Drag*, pub. by author, New Jersey, USA.
4.4 Baker, W. E. *et al.* (1983) *Explosion Hazards and Evaluation*, Elsevier, New York.
4.5 Glasstone, S. and Donlan, P. J. (1977) *The Effects of Nuclear Weapons*, 3rd edn, US Depts of Defense and Energy.
4.6 Christopherson, D. G. (1946) *Structural Defence (1945)*, UK Ministry of Home Security, Civil Defence Research Committee paper RC 450.
4.7 Mellsen, S. B. (1973) *Effect of Shielding on Blast Loading of Circular Cylinders*, Defence Research Establishment, Suffield, Alberta, Canada, Memorandum No. 13/72, February.
4.8 Newmark, N. M. (1984) Opening address, *Proc. Symp. on Soil-Structure Interaction*, Univ. of Arizona, Tucson.
4.9 Newmark, N. M. and Haltiwanger, J. D. (1962) *Air Force Design Manual: Principles and Practices for Design of Hardened Structures*, USAF Special Weapons Center, Kirkland AFB, Report AFSWC-TDR-62-138.
4.10 Newmark, N. M. and Hall, W. J. (1959) *Preliminary Design Methods for Underground Protective Structures*, USAF Special Weapons Center, Kirkland AFB, Report AFSWC-TR-60-5.
4.11 Hendron, A. J. and Auld, H. E. (1967) The effect of soil properties on the attenuation of air blast induced ground motions, *Proc. Symp. on Wave Propagation and Dynamic Properties of Earth Materials*, Univ. of New Mexico, Albuquerque.
4.12 Allgood, J. (1972) *Summary of Soil-structure Interaction*, US Naval Civil Engineering Laboratory, Tech. Rep. R. 771.
4.13 Bulson, P. S. (1985) *Buried Structures – Static and Dynamic Strength*, Chapman & Hall, London.

4.14 Bulson, P. S. (1967) Dynamic loading of buried tubes in sand and clay, *Proc. Symp. on Wave Propagation and Dynamic Properties of Earth Materials*, Univ. of New Mexico, Albuquerque.

4.15 *Domestic Nuclear Shelters* (1981) UK Home Office technical guide.

4.16 Schneider, W. G., Wilson, E. B. and Cross, P. E. (1946) Underwater explosives and explosions, *Effects of Impact and Explosion*, National Defense Research Committee, Washington, DC, USA.

4.17 Taylor, G. I. (1941) *The Pressure and Impulse of Submarine Explosion Waves on Plates*, UK Ministry of Home Security paper RC 235, July.

5

Structural loading from local explosions

5.1 CONCENTRATED EXTERNAL EXPLOSIVE LOADS ON SURFACE STRUCTURES

When an explosive charge is detonated in contact with, or very close to a structure in air, the loading can no longer be considered uniform over the area of component faces, and it is necessary to examine the spread and decay of the instantaneous pressure pulse as it travels across the external areas of the structure.

Local loading due to a contact explosion on a rigid surface in air has been summarized by Henrych [5.1], who considered first a prismatic charge, width b, length l, and height H in contact with its plan area bl. When detonated the products of the explosion are assumed to expand in the 5 directions normal to the external faces of the charge, and the instantaneous pressure on the target is equal at all points over the contact area and applied normally to the rigid surface. At a point on the surface distance x from the nearest face of the charge the time of arrival of the shock front will be x/v_s, where v_s is the speed at which the surface of the charge is displaced by the explosion. The impulse per unit area acting at a distance x is given by

$$\frac{I}{\text{area}} = x\rho u_x,$$

(5.1)

where ρ is the mass density of the explosive and u_x is the outburst speed.

The total contact explosion impulse is equal to the area of the impulse pattern, and for a prismatic charge with $b > 2H$ this is:

$$I = b^2 H\rho u_x \left(1 - \frac{2H}{b} + \frac{4}{3}\frac{H^2}{b^2}\right).$$

(5.2)

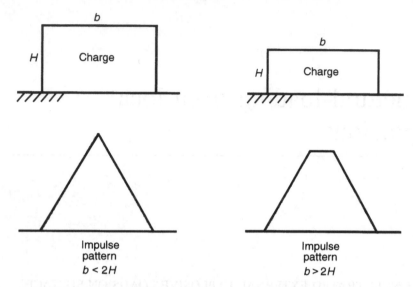

Figure 5.1 Triangular and trapezoidal impulse patterns (from Henrych, ref. 5.1).

The trapezoidal impulse distribution is shown in Figure 5.1. If $b < 2H$ the impulse distribution is triangular, and

$$I = b^2 H \rho u_x b / 6H. \tag{5.3}$$

Eq. (5.2) and (5.3) are also valid for a cylindrical charge placed with its base in contact with the rigid surface. For a hemispherical charge of radius r,

$$I = \frac{2\pi}{3} r^3 \rho u_x \cdot \tfrac{1}{2}, \tag{5.4}$$

and for a very flat charge having $H \ll b$,

$$I = blH\rho u_x. \tag{5.5}$$

In applying the above formulae care must be taken over units. The value of the outburst speed, u_x, is approximately equal to $(2Q)^{1/2}$, where Q is the work done when the explosive gases expand from their original volume to an infinite volume.

When the explosion occurs at a short distance from a rigid obstacle the pattern shape changes. Consider, for an example, a spherical charge of radius r exploding at a distance a from a flat surface, where $a < 10_r$. At a time t after the detonation the outburst pattern will be spherical, and as Figure 5.2 shows, reflection will be occurring at the same time that the central region of the explosion is still stationary. If the velocity of the outflowing particles (outburst

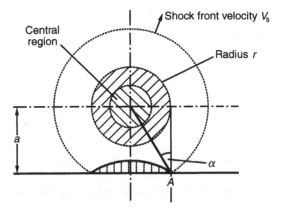

Figure 5.2 Spherical charge exploding close to flat surface (from Henrych, ref. 5.1).

speed) is u_x, as before, Henrych shows that the specific impulse, i, at point A, is

$$i = \left(\frac{A_0 W}{a^2}\right) \cos^4 \alpha, \tag{5.6}$$

where W is charge weight, $A_0 = (V_s + u_x)/4\pi$, and α is the angle between the vertical at A and the line joining A to the centre of the charge. V_s is the displacement velocity of the outburst surface.

For a cylindrical charge,

$$i = \left(\frac{2A_0 W_c}{a}\right) \cos^3 \alpha, \tag{5.7}$$

where W_c is the mass per unit length of the charge.

In order to find the total impulse on the surface it is necessary to integrate the specific impulse over the whole area. If the surface is a circular plate, the extremities of which have a value of α designated α_0, then

$$I = \pi A_0 W \sin^2 \alpha_0. \tag{5.8}$$

If the plate has infinite dimensions, so that α_0 approaches $\pi/2$, then

$$I = \pi A_0 W. \tag{5.9}$$

To use all the above equations we have to know the value of V_s and u_x. As discussed above, $u_x = (2Q)^{1/2}$, and V_s is the velocity of the shock front. From experiments it has been reported by Henrych that V_s has the following values in metres/sec for TNT and PETN: 7100 and 8450. The values given for u_x for the same two explosives are 6450 and 7700 metres/sec. Care must be taken when applying the above analysis that the units are correct.

Explosive charges close to the ground are often assembled in an array, to distribute total explosive mass over a large area and give some uniformity of

Structural loading from local explosions

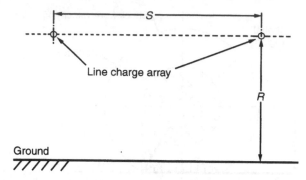

Figure 5.3 Schematic view of part of a line charge array (from Baker *et al.*, ref. 5.2).

surface blast loading. This type of arrangement has a military use connected with the rapid clearance of a lane through a minefield. Experiments on the blast output from a pair of parallel Primacord line charge arrays have been reported by Baker *et al.* [5.2], in which reflected pressures and ground impulses were measured at various distances from the arrays. Relationships between charge weight, line spacing and stand-off distance were analysed and scaling laws developed. In Figure 5.3, taken from the report, two identical line charges are shown, distance R above the ground, with a charge spacing of S. The model test conditions ranged from $S = 1.5$ in, $R = 2$ in, using 40 gram/foot Primacord, to $S = 9$ in, $R = 16$ in using 400 gram/foot Primacord. The greatest loads in the region below the charges occur when a Mach stem forms and charge spacings and stand-offs to achieve this formation were measured. The results are summarized in Figure 5.4, which compares scaled stand-off distance $(R/W^{1/3})$ and scaled charge spacing $(S/W^{1/3})$ at which a fully formed Mach stem is produced. At lesser stand-offs or spacing the pressure/time curve at ground level exhibits a

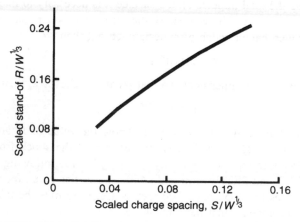

Figure 5.4 Position of parallel line charges to produce fully formed Mach stem on the ground surface (from Baker *et al.*, ref. 5.2).

'double peak' with the instantaneous incident pressure pulse followed by a separate reflected pressure pulse. Figure 5.4 suggests an approximate linear relationship of $S/R = 0.5$. Scaled reflected pressure curves show that reflected pressure increases very quickly when $R/W^{1/3} < 0.1$ and scaled reflected impulse curves show a rapid increase in impulse when $R/W^{1/3} < 0.07$.

5.2 GROUND SHOCK FROM CONCENTRATED UNDERGROUND EXPLOSIONS

The measurement of ground shock from the detonations of charges below the surface of soil was the subject of a large number of experimental programmes in the USA in the 35 years between 1950 and 1985, and over 100 tests were performed during this period. We have already discussed the Second World War tests in the USA reported by Lampson [2.12] in Chapter 2, but much more information has become available since then on loading, and on the magnitude and time histories of ground motions and stresses. Tests were reported in 1952 by Vaile [5.3] and in 1955 by Sachs and Swift [5.4], and since then a range of charge weights from 1 pound to 500 tons was used to investigate ground shock in a series of tests carried out by the Sandia Laboratory and the US Army Waterways Experiment Station in the USA. These were reported in a number of technical reports but were summarized and reviewed in a paper by Drake and Little [5.5] in 1983.

The ground shock produced by bombs or missiles exploding close to underground structures is an important threat. As we saw earlier, the stresses from underground explosions can be larger and have a longer duration than stresses resulting from similar bursts in air. The properties of soils and the presence of water have a noticeable effect on behaviour, but as Drake and Little pointed out, caution must be used in generalizing properties such as seismic velocity in any analysis. They summarized peak stresses in various types of soil in terms of a scaled range $R/W^{1/3}$ in ft/lb$^{1/3}$, as indicated in Figure 5.5, and then gave a number of empirical relationships that described ground shock data from the sources discussed above.

Stress and particle velocity pulses are characterized by exponential-type relationships that decay rapidly as outward propagation takes place from the centre of the explosion. The arrival time of the shock, t_a, is equal to R/c, where R is the distance from the explosion and c the seismic velocity for the soil. The shock rises to a peak with a rise time (t_r) of about $t_a/10$, and decays to zero in a time of t_a to $3t_a$. The stress in the soil, p, is given by

$$p = p_i\, e^{-\alpha t/t_a}, \qquad (t > 0), \tag{5.10}$$

where α is a time constant. The particle velocity, V, is given by

$$V = V_0(1 - \beta t/t_a)\, e^{-\beta t/t_a} \qquad (t > 0), \tag{5.11}$$

where β is a time constant, and ρ is mass density. Also,

$$V_0 = p_i/(\rho c), \qquad (5.12)$$

and experiments show that $\alpha \simeq 1.0$ and $\beta \simeq 0.4$.

Free-field stresses and ground motions for charges exploding close to a structure in soil are given by the following expressions:

$$\left.\begin{array}{l} p_i = f\rho c 160(R/W^{1/3})^{-n}, \\[6pt] V_0 = f160(R/W^{1/3})^{-n}, \\[6pt] I_0 = f\rho 1 \cdot 1(R/W^{1/3})^{1-n}, \end{array}\right\} \qquad (5.13)$$

where I_0 is peak impulse, n is an attenuation coefficient and f a factor that depends on the depth below the surface of the charge, and is sometimes known as the ground shock coupling factor. It is the ratio of the ground shock

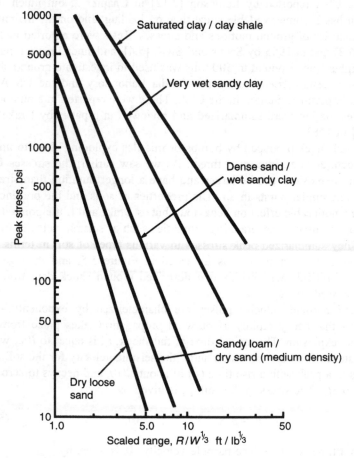

Figure 5.5 Peak stresses for underground explosions in various types of soil (from Drake and Little, ref. 5.5).

Table 5.1

Scaled depth of explosion $(d/W^{1/3})$ in ft/lb$^{1/3}$	0	0.2	0.4	0.6	0.8	1.0	1.2	1.4
Ground shock coupling factor, f		0.4	0.6	0.7	0.8	0.88	0.92	0.97	1.0

magnitude at shallow depths of burial to the magnitude at deep covers. Values given for f are listed in Table 5.1.

For design purposes Drake and Little gave the values for seismic velocity (c), Acoustic Impedance (ρc) and attenuation coefficient (n) given in Table 5.2.

In fact, the attenuation coefficient, n, can be estimated directly from seismic velocity, as given in Table 5.3.

It was also noted that
    : peak accelerations were proportional to seismic velocity
    : peak displacement was inversely proportional to seismic velocity
    : peak impulse was sensitive to density variation;

and that near the source of the explosion peak particle velocities in soil (V_0) were virtually independent of soil properties.

A large number of ground shock experiments were monitored by the South West Research Institute, Texas, USA, in the early 1980s, and a summary of the findings was reported by Westine and Friesenhahn in 1983 [5.6]. The buried detonations were from the explosion of charges in mortar and artillery projectiles. The authors divided their analysis of results into unsaturated and saturated soils, and developed empirical equations for the relationship between scaled maximum pressures and the scaled distance from the explosion.

Table 5.2

	c ft/sec	ρc psi/ft/sec	n
Loose dry sands and gravels with low density	600	12	3
Sandy loam, dry sands and backfill	1000	22	2.75
Dense sand, high density	1600	44	2.5
Wet sandy clay with air voids >4%	1800	44	2.5
Saturated sandy clay and sands with small amounts of air voids (<1%)	5000	48	2.5
Heavy saturated clays and clay shales	>5000	150	1.5

Table 5.3

Seismic velocity (c) in ft/sec	500–600	750–1000	1000–1400	1400–1800	>5000
Attenuation coefficient (n)	3–3.5	3	2.75	2.5	1.5–2.25

The general empirical equation for predicting free-field ground shock pressure from the detonations of buried explosive was:

$$(p_{max}/\rho c^2)(4.35 + y/d)^{-1}(0.25+0.75 \tanh(0.48\rho^{1/3}c^{2/3}d/E^{1/3}))$$
$$= 0.0175(E^{1/3}/(\rho^{1/3}c^{2/3}R_e))^{3.42}, \qquad (5.14)$$

where p_{max} = maximum pressure, ρ = mass density of the soil, c = seismic velocity for soil, d = depth of explosion below ground surface, y = depth of point at which the pressure is required, E = energy release of the explosive, and R_e = effective slant range between the point under consideration and the centre of the explosion, taking account of the charge geometry and orientation.

R_e was given by

$$R_e/l = (M - N^2 - 0.25)^{1/3}$$
$$\times [(N + 0.5)/(M + N)^{1/2} - (N - 0.5)/(M - N)^{1/2}]^{-1/3}, \qquad (5.15)$$

where

$$M = (z/l)^2 + (x/l)^2 + (y/l)^2 + 0.25,$$
$$N = (y/l) \cos \theta + (z/l) \sin \theta,$$

and

z = horizontal distance of point under consideration to the vertical plane in which the length of the charge lies

x = transverse distance of the point under consideration from the centre of the charge

l = length of explosive charge

θ = angle of orientation of the bomb ($\theta = 0°$ is vertical)

The format of Eq. (5.15) allows scaled pressure ($p/\rho c^2$), which is non-dimensional, to be plotted against scaled effective stand-off distance ($R\rho^{1/3}c^{2/3}/E^{1/3}$), which is also non-dimensional, since the units of energy (E) are force × length. Figure 5.6 shows a plot of Eq. (5.15) in terms of scaled pressure and scaled effective stand-off, with parallel lines indicating the spread with one standard deviation for normal distribution of a large number of test results conducted at the South West Research Institute. This plot also gave very good predictions of the scatter of experimental results from tests with mortar and artillery shells at the Waterways Experiment Station and with C-4 explosive charges fired at White Sands, USA. These tests were carried out in silt and sandy soils, well clear of any water table. When Figure 5.6 was compared with test results for 500 lb and 2000 lb bombs carried out at Fort Knox, USA, in saturated soft brown clay with the water table at a depth of 10 feet, the scaled pressures were an order of magnitude greater than would be predicted by Figure 5.6. This was mainly because the seismic velocity in soft clay of 1200 ft/sec was increased to the speed of sound in water (4800 ft/sec). In this instance it is more accurate to use data we discussed earlier from the work of Cole [2.22], to give the following equation for shock pressures in water:

$$p = 24\ 650 \left[\frac{W^{1/3}(lb \cdot \text{TNT})^{1/3}}{R(\text{feet})} \right], \quad psi \qquad (5.16)$$

which can be rewritten as:

$$p/4.35\rho c^2 = 0.04224(\rho^{1/3}c^{2/3}R/W^{1/3})^{-1.16}. \qquad (5.17)$$

This equation gives good agreement with the Fort Knox tests, and as we might expect indicates that free-field ground shock pressures are considerably influenced by the state of saturation of soils.

All the experimental results depend on the accurate measurement of free-field stresses, and measurements are influenced by the degree to which buried stress gauges disturb the soil in their immediate vicinity. The influence of gauge

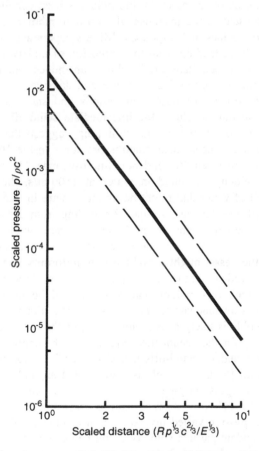

Figure 5.6 Scaled pressure v scaled effective stand-off distance from tests on bombs, shells and charges (from Westine and Friesenhahn, ref. 5.6).

installation procedures in rock and soil has been discussed by Florence, Keough and Mak [5.7]. In particular they investigated theoretically the effect of bonded and unbonded interfaces between the gauge and the surrounding soil. They assumed that a bonded interface would transmit compression and shear, and an unbonded interface compressive loads only. Their calculations show that lack of interface bonding produces a non-uniform stress distribution at the gauge which produces inaccuracies at low stresses. At higher stresses plastic flow in the soil increases, and causes the stress distribution to be more uniform. This improves accuracy.

5.3 CONCENTRATED EXTERNAL LOADS ON UNDERWATER STRUCTURES

Much of the load-measuring research in this field has been conducted by naval research teams interested in the response of ship and submarine structures to local explosions from mines and torpedoes. Minesweepers and similar vessels are likely to suffer shock loading from the explosion of underwater charges at what are called 'intermediate stand-offs', which can induce flexural motion or 'whipping' of the hull as well as local rupture and damage. We have already examined the characteristics of bubble pulsations and shock waves at long distance, but now we must evaluate the intermediate stand-off effects. Hicks [5.8], in discussing explosion-induced hull whipping, suggests that British naval scientists are interested in peak instantaneous pressures in the 100 to 400 bar range, with durations between 300 and 600 microseconds. He concludes that with the acoustic velocity of sound in water at 1500 m/s, the shock wave occupies a thin shell of water that might be about 1 metre in thickness.

Hicks investigated how far the formulae for loading from distant explosions can be applied to close explosions. In the latter case there would seem to be a loss of accuracy for explosions near the bow and stern, where divergent flow around the end of the vessel might introduce non-uniformity into the analysis. Hicks, however, concluded that the accuracy of the distant-flow approximations for close point sources was extraordinarily good, and he assumed that the distant-flow analysis was adequate for normal stand-offs. For very close stand-offs the pressure bubble is likely to completely engulf the hull and this distorts the analysis too much for the distant-flow equations to be applicable.

Loading from intermediate stand-offs has also been investigated recently by Haxton and Haywood [5.9], who took as their model an underwater cylinder subjected to a pressure pulse originating from a point distant about one radius from the wall of the cylinder. It was assumed that the peak instantaneous pressure from the explosion varied inversely with distance, and that the radial component of the incident particle velocity could be neglected in the shadow of the cylinder. Referring to Figure 5.7, the peak instantaneous pressure depends on the radius R_t and the angle θ_t, and the point of grazing incidence, beyond which

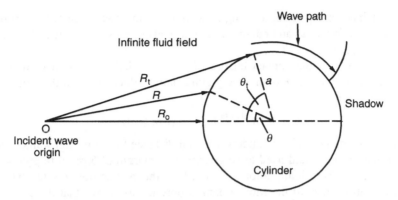

Figure 5.7 Underwater cylinder subjected to pressure pulse (from Haxton and Haywood, ref. 5.9).

the wave path length must be measured around the surface of the cylinder. The following equations were derived for p_1:

$$p_1 = \frac{2R_0}{\pi \varepsilon_n} p_0 \, e^{-(t+R_0/c)/T} \int_0^\theta \frac{e^{R/Tc}}{R} \cos n\theta d\theta,$$

$$0 \le \theta \le \theta_t,$$

and

$$p_1 = \frac{2R_0}{\pi \varepsilon_n} p_0 \, e^{-(t+R_0/c)/T} \cdot \left[\int_0^{\theta_t} \frac{e^{R/Tc}}{R} \cos n\theta d\theta + \int_t^\theta \frac{e^{R'/Tc}}{R'} \cos n\theta d\theta, \right.$$

$$\theta_t < \theta < \pi$$

$$\left. \right\} \quad (5.18)$$

where

$$R = [a^2 + (R_0 + a)^2 - 2a(R_0 + a) \cos \theta]^{1/2},$$

and

$$R' + [R_t + a(\theta - \theta_t)].$$

T is the decay time constant, c is the velocity of sound in water, $\varepsilon_n = 2$ when $n = 0$ and $\varepsilon_n = 1$ when $n \ge 1$ where n is the circumferential modal index for the deflections of the cylinder wall.

A long time before analytical work of the above type was attempted, the problems of close-in explosions had been discussed by Christopherson [4.6]. In 1945 no satisfactory measurements had been made of the pressure inside a gas bubble, and when an underwater target was very near the source of the explosion there was no general theory of loading. There had been tests during the Second World War at the UK Road Research Laboratory which suggested that a charge of TNT of weight W lb in contact with a lightly reinforced concrete slab of thickness t ft underwater would produce heavy damage (severe cracking and bowing) if $t/W^{1/3} < 1.33$. This compares with a value of $t/W^{1/3} < 1.1$ for a charge confined in the earth, and suggests that the underwater charge is about 1.75 times as effective as an underground charge against a similar target, for this type of damage.

It was found that the contact charge of TNT needed to cause a complete breach in a masonry or mass-concrete gravity dam of thickness t, at the point of attack, was given by

$$W = 6.10^{-3}t^{3.75} \text{ lb}, \qquad (t \text{ in feet}) \qquad (5.19)$$

and that if the charge exploded underwater at a distance from the dam wall equal to t then the charge would need to be greater by a margin of 5 or 10 to produce similar damage. This information was used in the preparation for the well-known operations against the Möhne dam, where it was shown that the experimental predictions at small and medium scales were verified at full scale.

5.4 BLAST LOADS IN TUNNELS AND SHAFTS

Looking first at the history of research, we note that E. B. Philip reported in 1944 [5.10] on the propagation of air blast through systems of tunnels. Much of the research was empirical, using piezo-electric gauges, and was undertaken to assess blast wave decay in underground workings and underground railways which were used during the Second World War as air-raid shelters or bomb-proof factories.

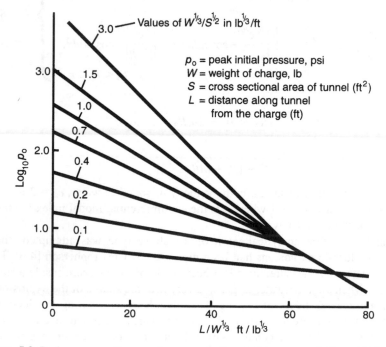

Figure 5.8 Peak pressure from a charge exploding inside a tunnel (from Philip, ref. 5.10).

There were two problems, the first resulting from an explosion in the tunnel, the second due to the entry of blast from an explosion outside. A selection of Philip's results for explosions within a long straight tunnel was given by Christopherson [4.6] and some are shown in Figure 5.8. They refer to brick-lined tunnels, in which pressure decay is likely to be quicker than in smoother tunnels of concrete or steel. The rate at which the peak pressures decayed was influenced by the parameter $W^{1/3}/S^{1/2}$, where S is the cross-sectional area in square feet, and W is the weight of the charge of TNT in lb. Although the peak pressures gradually decayed along the tunnel it was interesting to note that at short distances from the explosion the blast impulse had increased slightly, and for very small charges this increase could still be measured at long distances from the point of explosion, as indicated in Figure 5.9.

It was clear that in long straight tunnels the decay was slow enough to suggest that damage due to flying debris might occur at long distances from the point of the explosion, and this led to the policy of interrupting long straight runs with 'blast traps'. Christopherson reported estimates of the effect of a variety of trap shapes based on trials at the Road Research Laboratory, and a selection of these estimates is shown in Figure 5.10. The factor K_p should be applied to the pressures in straight tunnels given in Figure 5.8, if a trap of the geometry indicated is interposed between the charge and the point under consideration. The factor K_2 gives similar information for impulse, and is applied to the impulses from Figure 5.9. The estimates from the Road Research Laboratory suggested that a simple right-angle bend reduced the peak pressure by 30%, and

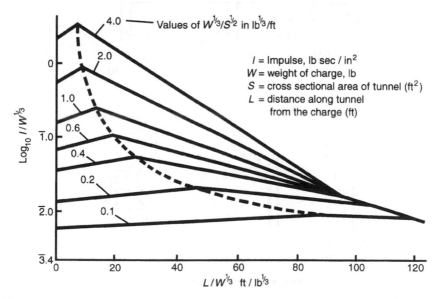

Figure 5.9 Blast impulse from a charge exploding inside a tunnel (from Philip, ref. 5.10).

a uniform diameter tee junction by about 50%. The reduction in a tee can be increased slightly if the branching pipe that forms the stem of the tee has either half or double the cross-sectional area of the main pipe.

When figures were given in 1965 in ref. [1.10], there were some discrepancies. In this reference the reduction at a simple right-angle bend was given as 50%. The system of four right-angle bends shown in Figure 5.11 was said to reduce the pressure by 85%, and if extensions or pockets were introduced at the

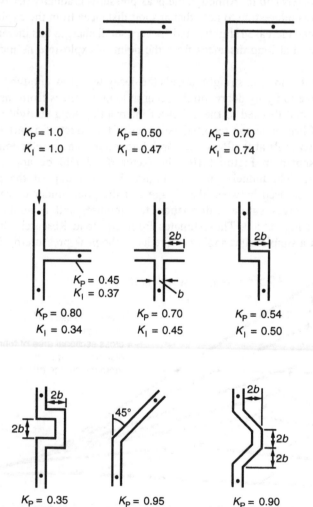

All tunnels have a diameter of *b*

Figure 5.10 Comparative reduction in maximum positive blast pressure and positive blast impulse, K_p-pressure, K_i-impulse (from Christopherson, ref. 4.6).

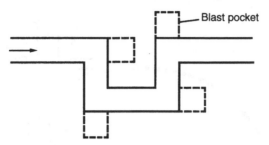

Figure 5.11 Four right angle bends with blast pockets (from Christopherson, ref. 4.6).

bends (shown dotted), the reduction was 92%. The Road Research Laboratory figures were 65% and 77%, but the latter was increased to 86% if the pocket depth was double the pipe diameter. The introduction of a rectangular pipe system (Figure 5.12) was said by the US Army Technical Manual to lead to a reduction of 85%, whereas the Road Research Laboratory figure was 62%.

Later research suggested that the above figures were over optimistic, and more recent publications in the USA and elsewhere propose more conservative design rules. Much of this information still seems to be restricted, but open recommendations are given in a book by Henrych [5.1] published in English in 1979. He gives information on the propagation of a shock wave in cranked and branched channels in which he notes that the factor K_p changes as the peak entering pressure increases. Figure 5.13, taken from [5.1], gives a range of values of K_p for several branch geometries. At the higher pressures the reduction at a simple right angle bend is only 10%, so that $K_p = 0.9$. For successive right-angle bends his results show that the same coefficient is applied each time, so that for three successive bends the overall value of K_p is $0.9^3 = 0.73$. His results suggest that for n successive bends the overall factor is 0.9^n, but there is no analytical proof of this. What is very apparent from a study of all the experimental information is that the smoothness of the inside of the tunnels has an important bearing on the results, and as this information is not always given in experimental reports it is dangerous to correlate the conclusions of separate

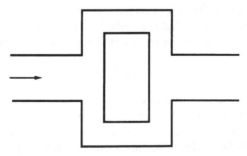

Figure 5.12 Rectangular pipe system blast trap (from *Fundamentals of Protective Design*, TM5-855-1, 1965).

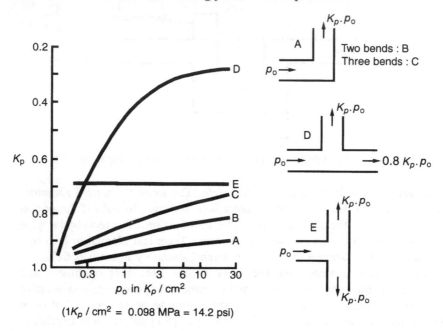

Figure 5.13 Propagation of a shock wave in branched channels (from Henrych, ref. 5.1).

experimenters. We can perhaps assume that the low values of K_p given in the UK data of the 1940s were measured in tunnels with very rough walls, whereas later and more conservative results from the USA and Europe were gathered from smooth-walled tunnels. The 1986 edition of the US Army Technical Manual TM-855-1 gives $K_p = 0.94''$, when the initial pressure (p_0) $\leq$ 50 psi.

The entry of blast from a 'free field' into a side-on tunnel often needs to be assessed for design purposes. When a pressure pulse having a peak initial pressure of p_0 passes transversely over a side-on tunnel, the actions are similar to those for the T-branch tunnels given at D in Figure 5.13. At low values of p_0, $K_p = 1$, but at higher values K_p is reduced to about 0.3. This value of K_p is probably a little unconservative, but is satisfactory for preliminary calculations. Curve D in Figure 5.13 can also be represented approximately by the equation

$$K_p = n(p_0)^n. \tag{5.20}$$

When the centre of the explosion is directly in line with the mouth of the tunnel, as shown in Figure 5.14, Henrych quotes experiments which show that it is sufficient to substitute αR for the distance R of the entrance from the explosion centre, in order to determine the value of the peak pressure p_0 in the entrance of the tunnel (assuming no reflecting surface around the inlet opening). The relationship between α and $\log p_0$, where p_0 is in kg/cm^2, is shown in Figure

5.14. The distance αR is substituted for R in calculating z, used in expressions relating pressure to scaled distance, as in Eq. (1.3) for example.

The consequences of nuclear blast entering underground railway systems were assessed by a number of test programmes in the UK and USA during the 1960s, and the results are still in limited circulation. However, Henrych has quoted experimental studies on the decay of peak pressure along a pipeline due to the roughness of the surface. He suggests that the pressure decrease is exponential and that at a distance x along the tunnel, the peak pressure p_x is given by

$$p_x = \frac{p_0 \alpha R}{(\alpha R + x)} e^{-0.4f(x/r)},$$ (5.21)

where f is the hydraulic coefficient of friction, r is the pipe radius, and there is no reflecting surface at the inlet opening. The value of f can be found from the approximate formula

$$f = (2 \log r/2h + 1.74)^{-2},$$ (5.22)

where h is the height of the roughness.

From the work of Henrych it is clear that in comparison with 'free field' conditions, the peak instantaneous pressure at a given range along the tunnel is much less than would be measured in air at the same range from an explosion,

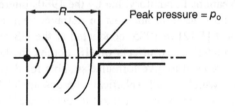

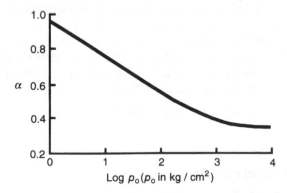

Figure 5.14 Entry of blast into a tunnel from an explosion in line with the tunnel mouth (from Henrych, ref. 5.1).

that durations of the positive phase tend to be longer, and that the overall effect on impulse is to reduce it by a rough figure of 50% at any range.

A review of blast wave behaviour in tunnels was reported in 1968 by Taylor [5.11], in which he made the point that there is a staged progression of a shock wave moving through a 90° turn in a tube. The expanding shock will produce different pressures at all points across a cross section drawn one-half diameter downstream from the bend; then the blind wall reflection forms and moves across the same cross section to produce a second jump in pressure. Measurements in a shock tube have shown that pressure fluctuations one diameter downstream from a 90° turn can exceed the input pressure; in other words a turn amplifies pressure locally before attenuating it. Taylor shows the measurements of peak pressure at 2 and 28 diameters downstream from the junction, and indicates that at 2 diameters the peak pressure is higher than at 28 diameters. Downstream shock pressure measurements at Y and T junctions were also reported by Taylor, and pressures at 10 diameters downstream were given by him. As we saw earlier, pressures were not halved when the routes are doubled.

If the diameter of a straight tunnel is suddenly increased there will be a reduction in peak pressure. Relationships were given for incident and transmitted overpressures for area ratios of 7.3, 18.5 and 66.

Most of the experimental work of the 1960s was carried out in relation to explosions of long duration by the US Army Ballistic Research Laboratory and the US Air Force Weapons Laboratory, but by the 1980s interest had returned to the attenuation of short duration air blast in entranceways and tunnels. Tests were reported by Britt [5.12] in 1985, conducted at the US Waterways Experiment Station, that modelled the effects of HE bombs in the 100 to 1000 kg range. The results were used to give formulae suitable for design. Some 99 tests were conducted, in which small spherical charges of C-4 explosive were detonated outside and inside the entrance to tunnels of circular and square cross section. The model tunnels were 30 cm diameter and between 4 and 24 diameters long; blast pressures were measured along the tunnels when charges were detonated from end-on and side-on positions relative to the tunnel mouths. The results were compared with earlier work on tunnel pressures in which the pressures were relatively much greater – almost by an order of magnitude. These were reported in 1977 by Itschner and Anet [5.13]. Other earlier test results from Germany had been given by Gurke and Scheklinski-Gluck in 1980 [5.14].

The peak pressure attenuation was given by the equation

$$p_x = p_i/(1 + \tan[(\pi/2)\bar{x}/(\bar{x} + \bar{E})]),\qquad(5.23)$$

where

p_x = pressure at a distance x down the tunnel

p_i = pressure at tunnel entrance

$\bar{x} = x/A^{1/2}$, where A is the cross-sectional area of a circular or square tunnel

$\bar{E}$ = dimensionless decay parameter.

For end on bursts (i.e. in line with the entrance to the tunnel), $\bar{E}$ was given by

$$\bar{E} = K_e(W^{2/3}/A)(p_0/p_i)^{0.4}, \tag{5.24}$$

where

W = charge weight, p_0 = atmospheric pressure

K_e = 0.586 if W is in kilograms and A in square metres, or

K_e = 3.72 if W is in pounds and A in square feet.

For side-on bursts, in which the charge explodes at a lateral distance from the entrance to the tunnel, $\bar{E}$ was given tentatively as

$$\bar{E} = K_s(W^{2/3}/A)(p_0/p_i)^{0.8}, \tag{5.25}$$

where

K_s = 2.26 for W in kilograms and A in square metres, or

14.4 for W in pounds and A in square feet,

but further experimental work was said to be needed to confirm this.

These equations were stated by the author to be valid within the following ranges:

$$1 < W^{1/3}/A^{1/2} < 4 \text{ kg}^{1/3}/\text{m}; \qquad 0.3 \text{ MPa} < p_i < 40 \text{ MPa (end-on)};$$
$$0.6 \text{ MPa} < p_i < 11 \text{ MPa (side-on)}$$

Britt also examined values of impulse, which was virtually constant along the tunnel, and he gave the following relationships,

$$I = I_{0E} W^{1/3}(W^{1/3}/A^{1/2})^{1/2}(p_i/p_0)^{0.6} \text{ (end-on)} \tag{5.26}$$

where

I_{0E} = 0.042 for I in $MPa \cdot$ msec, W in kg and A in m^2,

or

I_{0E} = 7.43 for I in psi $\cdot$ msec, W in lb and A in ft^2.

For side-on bursts

$$I = I_{0s} W^{1/3}(W^{1/3}/A^{1/2})^{1/2}(p_i/p_0)^{0.2}, \tag{5.27}$$

where I_{0s} = 0.0756 and 13.4 respectively.

A more detailed discussion of this experimental study is given in ref. [5.12].

One of the main areas of interest in the 1980s was the development of computer codes to calculate shock wave propagation in systems of air ducts. The source of the shock was taken to be a surface or air nuclear explosion at a distance, or a conventional high explosive detonation within the ducting system.

These codes enable extremely complex systems with up to 1000 branches to be analysed for a very wide range of inlet geometries and bomb orientations. Codes of this type can be used, for example, to model the effect of shock waves in the ducting system of aircraft or ship structures, and are valuable tools in analysing the strength of such structures under attack by terrorist explosive devices. Most of the codes are marketed by the developers, and a comparison of their capabilities is beyond the purpose of this book. However, one code which was commissioned by the US Army Corps of Engineers, Omaha District, and written at the University of Nevada, has been described in some detail by Fashbaugh [5.15]. The code, which can be used on a PC computer, enables the shock propagation in ducting systems (SPIDS) to be calculated sequentially. The shock propagation in a main duct, for example, is calculated and the time history of pressures and temperatures is stored on a computer file. This file then specifies pressures and temperatures at the inlet of a branch duct from which the shock propagation in the duct can be deduced. This procedure can be repeated for large numbers of branches – up to ten for each specific duct. There are limitations, one of which is that reflected shock waves from branch ducts into main ducts are not accounted for; neither are shock wave losses as the wave passes the entrance to a branch duct.

Simulation is achieved by evaluating flow losses from a point s just outside the duct entrance to a point e inside the entrance, using the collected results of many experimental programmes. The duct entrance static pressure p_e is related to the pressure outside the duct, p_s, by the relationship

$$p_e = p_s \left(\frac{p_{te}}{p_{ts}} \right) \left[\left(\frac{T_e}{T_{te}} \right) \left(\frac{T_{ts}}{T_s} \right) \right]^{\gamma/(\gamma-1)}, \tag{5.28}$$

where p_{te}/p_{ts} is the stagnation pressure ratio, given by

$$\frac{p_{te}}{p_{ts}} = A(2) \, e^{-A(1)M_s}. \tag{5.29}$$

T_e and T_s are static temperatures at points e and s; T_{te} and T_{ts} are stagnation temperatures. $A(1)$ and $A(2)$ are empirical coefficients that depend on the entrance geometry and position of the explosion, and M_s is the flow Mach number outside the duct. The shock wave attenuation due to internal friction in the ducting is represented by a friction factor, given in terms of $(h/D)^{1/2}$, as discussed earlier, where h is the average wall protrusion height and D is the duct diameter. Fashbaugh used the equation

$$t = 0.016 + 0.00491(h/D)^{1/2} + 0.258(h/D), \tag{5.30}$$

and for a side-on entrance $A(1) = 1.1398$, $A(2) = 1.24$. The linear relationship between p_{te}/p_{ts} and M_s from experiments is shown in Figure 5.15. The code has been shown to give good agreement with tests for the decay of shock pressure with distance along a duct, and for the effects of a sudden increase in the cross-sectional area of a duct.

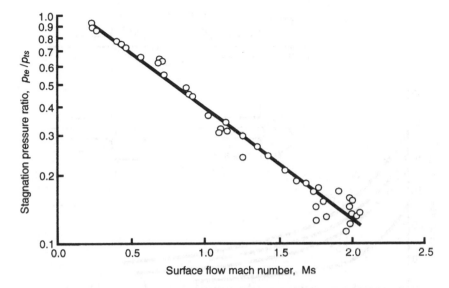

Figure 5.15 Relationship between the stagnation pressure ratio and the surface flow Mach number for side-on ducts (from Fashbaugh, ref. 5.15).

Recent experimental studies have also been conducted at the AC-Laboratorium at Spiez in Switzerland, and were reported in 1991 by Binggeli and Anet [5.16]. The test results were used to find the values of a and b in the equation

$$p_s = a(L/D)^{-b}, \qquad (5.31)$$

where p_s is the pressure at a distance L along the tunnel given by the parameter L/D, where D is the tunnel diameter. Values of the exponent b are functions of the weight of the explosive charge and the distance from the point of the explosion to the centre of the tunnel entrance. The coefficient a was found for a range of charge weights from 100 kg to 1500 kg. Typical results for a charge weight of 1500 kg set at various distances and angles from the tunnel mouth are given in Figure 5.16, in which the value of peak pressure at a distance $L/D = 2$ along the tunnel is shown for four values of the stand-off angle α (0°, 30°, 60° and 90°). Note that when the distance A to the charge = $4m$, the value of p_s changes with orientation according to Table 5.4 (given for a range of charge weights).

5.5 LOADS FROM CONFINED EXPLOSIONS IN UNVENTED AND VENTED CHAMBERS

If the detonation of a high explosive charge occurs in a space confined by infinitely strong inflexible and airtight boundaries there can be no escape of the gaseous products and no change in the enclosed volume. The shock wave will be

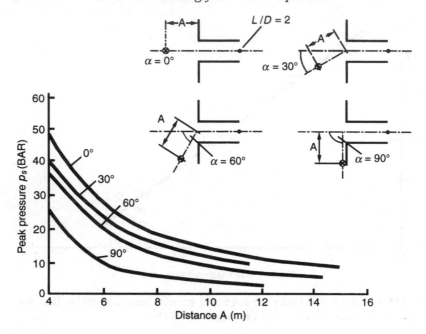

Figure 5.16 Relationship between peak pressure, distance A and stand-off angle (α) for a charge of 1500 kg, when L/D = 2 (from Binggeli and Anet, ref. 5.16).

reflected from all faces, and the reflections will continue until the energy of the explosion is expended in heat and perhaps by some form of absorption by the confining walls. If the raising of the internal temperature is the only way of expending energy the final equilibrium pressure will be higher than the original pressure within the space. This permanent pressure rise would be a gas pressure rather than an instantaneous peak pressure at a shock front.

An unyielding perfectly confined space is unusual. Most chambers are connected to the atmosphere by ducts or tunnels, or are deliberately vented to reduce the damaging effects of an internal explosion. Chambers in aluminium structures such as aircraft shells have boundaries that can deflect under pressure, so that the elasticity of the containing structure cannot be ignored. These and other problems meant that in the early days of research into confined explosions the exact analysis of blast pressure was very difficult, and at the end of the Second World War the recommended design loads for internal explosions in

Table 5.4

Weight	Stand-off angle	0°	30°	60°	90°
1500 kg	pressure p_s (bar)	50	40	37	25
1000 kg		30	26	22	18
500 kg		16	12	12	10

confined structures were very approximate. In 1945 Christopherson [4.6] recorded an empirical relationship for explosive pressures in a rigid enclosure with venting at one end, and we shall come to this later, but it was not until the 1960s and 1970s that the problems of internal blast loading were satisfactorily analysed for chambers of simple shape. Since then the development of computer codes has enabled the internal blast behaviour within very complex boundary geometries to be predicted.

The shock wave loading from an explosive charge within a chamber has been concisely described by Baker *et al.* [4.4], drawing on the earlier work of Gregory [5.17], Baker [5.18] and Kingery, Ewing and Schumacher [5.19]. The authors point out that the internal pressure loading consists of two phases, the first due to direct reflection waves from the internal walls plus reflected pulses that arrive at the walls. These would be attenuated in amplitude and be very complex in waveform. The initial maximum pressures can be found from data for the normal blast wave reflection from flat or curved surfaces, but oblique reflections can generate Mach waves and lead to a concentration of high pressures in corners of the chamber.

The second phase occurs when the shock waves that have been reflected inwards strengthen as they implode towards the centre of the chamber and reflect again to load the structure impulsively for a second time. Attenuation will occur so that eventually the shock pulses will be very small and will die out altogether. Figure 5.17 is an oft-quoted schematic representation first given by Gregory in 1976, showing shock reflections from the interior walls of a cylindrical containment structure. The loads for this geometry can be calculated by using two-dimensional computer programs, and the predictions compared with test results. More complex shapes were difficult to analyse at that time, and internal blast loading was usually found experimentally. Baker *et al.* [4.4] nevertheless investigated simplified theoretical loading for initial design purposes. They made three assumptions:

(a) incident and reflected blast pulses could be represented by triangular shapes with an abrupt rise;
(b) the duration of these pulses could be adjusted to preserve the correct impulses;
(c) initial internal blast loading parameters are always the normally reflected parameters, even for oblique reflections from the walls.

Assumption (c) is virtually correct up to the angle for limit of regular reflection, which we saw earlier was about 40° for strong waves and 70° for weak waves. For chambers approaching a cube in shape shock reflections from the walls will be regularly reflected in most parts of the cube. Approximate estimates of reflected wave magnitude were also made in ref. [4.4], where the second shock was taken to be half the amplitude of the initial reflected shock, and the third as half the amplitude of the second. Any further reflections were assumed to have insignificant amplitudes. Furthermore, it was assumed that the time delay between successive shock loads was equal to twice the time to the initial shock pulse. These approximations lead to a simplified pressure loading at a point on the inner surface of the chamber consisting of three successive triangular pulses with peak pressures p_r, $p_r/2$ and $p_r/4$, each associated with a duration of t_r, and the shocks occurring at intervals of t_a, $3t_A$ and $5t_A$.

The development of computer codes to describe the loading actions inside closed structures began in the early 1970s. It has been pointed out in a recent paper by Swisdak and Montanaro [5.20] that the original development in the field was due to Proctor [5.21] in 1972, when he produced a program for the US Naval authorities called INBLAS. This dealt originally with shock loading typical of TNT explosions, but his work was later amended by Ward and Lorenz to include deflagration loading typical of gas or vapour cloud explosions. In the early 1980s a significant improvement was made by Britt *et al.* [5.22], who replaced the original shock calculations with an analysis of greater accuracy and so formed the BLASTINW code.

This code was designed to run on a mainframe computer, but the rapid growth of desktop personal computers meant that it was worth updating and combining the better features in INBLAS and BLASTINW to form a revised code for the prediction of blast inside closed or vented structures. This code, now called INBLAST, requires the input of a table of pressure versus material consumption

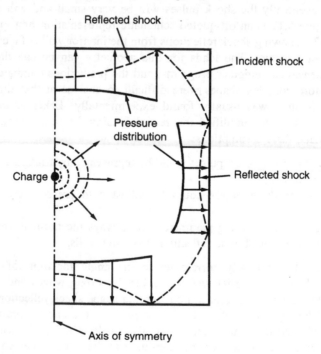

Figure 5.17 Shock reflections from interior walls of cylindrical containment structure (from Gregory, ref. 5.17).

rate, and a table of total burning area versus the weight of material consumed. With this information the relationship between pressure and time in all chambers of the structure can be calculated. The program uses techniques developed earlier for a low-altitude multi-burst code (LAMB), which is described under NATO restrictions by Britt and Drake in ref. [5.23], and which predicts the direct and multiple reflected shock waves present after a closed chamber detonation. INBLAST has been shown to predict accurately confined gas pressures as a function of loading density (charge weight/chamber volume), and a comparison of predictions with test results is shown in Figure 5.18, taken from ref. [5.20].

An earlier paper by Britt and Drake [5.24], presented without restriction in 1985, examined the propagation of short duration blast into chambers from high explosive charges detonated at the entrance to openings to the chambers. This work combined the ducting and tunnel technology with the chamber analysis, and resulted in a useful and progressive comparison between theory and experiment.

Immediately after the explosion the instantaneous increase in external pressure causes a high-velocity jet to enter the chamber through an opening of area A. Eventually the internal and external pressures become equal, and Kriebel [5.25] had shown in 1972 that the time in milliseconds was equal to $V/2A$ in

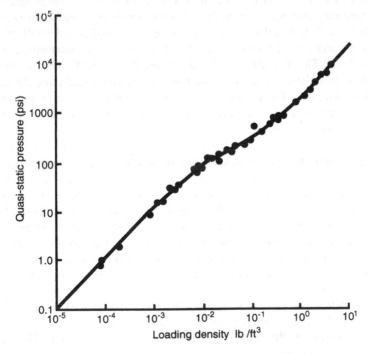

Figure 5.18 Comparison of measured quasi-static pressure with Inblast computations (from Swisdak and Montanaro, ref. 5.20).

units of feet, where V is the chamber volume. At about the same time shock tube experiments at the US Army Ballistics Research Laboratory were carried out by Coulter [5.26] in which chamber volumes, entrance areas, pressures and geometries were varied. It was shown that the pressure wave expanding into the chamber decays in amplitude and reflects from internal surfaces. The attenuation of the instantaneous peak pressure was inversely proportioned to the opening diameter, and its positive duration was directly proportional to the diameter. For non-circular openings the diameter was replaced by a 'mean opening dimension'. Tests at the US Army Waterways Experiment Station in the early 1980s, using C-4 and TNT charges, showed that side-on peak pressures in the test chambers (p_{max}) were related to the peak pressure in the opening (entrance tunnel) (p_0) by the equation

$$p_{max}/p_0 = 0.65(1 - 0.25\alpha)(R/D)^{-1.35}, \tag{5.32}$$

where R is the distance of the centre of the opening to the pressure gauge in the chamber and D is entrance tunnel diameter. α is the angle in radians between the normal to the centre of the opening and the gauge location. For small angles it was sufficient to ignore α and within the units of experimental scatter to simplify the above equation to $p_{max}/p_0 = 0.65(R/D)^{-1.35}$. The values of p_{max} are due to the first side-on peak at the gauge as the result of incident pressure and do not take account of internally reflected waves. If information about reflected waves is required, then a more complex analysis involving a shock diffraction model is needed. Path lengths of 'rays' of successively higher order reflections are generated, and arrival times calculated; then shock wave attenuation with distance is found. The procedure is fully described in ref. [5.24]. The combined pressure pulse in a chamber, found in this way, can be calculated using the code CHAMBER, originally for a mainframe computer but converted to run on a desktop computer. Predictions for the initial peak reflected shock were shown to agree well with test results.

It is clear that the levels of pressure from an internal explosion, whether due to a high-explosive detonation or a vapour or dust cloud deflagration, can be reduced by the judicious use of vents. The design of venting systems is governed by the pressure reductions that are required, so there has been a considerable research effort in this area. Going back to the work of Christopherson [4.6], which we mentioned earlier, we noted that he analysed the effect on the pressure in a rigid rectangular chamber if one end were vented. The 'partial enclosure' was the subject of an empirical relationship in which the results of a small number of observations were recorded in terms of the charge weight of TNT, the volume (V) in cubic feet, and the face-on blast impulse on the venting side wall of a cubical enclosure (A_T), in lb m sec/in^2. His results are plotted in Figure 5.19 on the axes $A_T/W^{1/3}$ and V/W (the volume of the chamber per lb of charge). For initial design purposes, the relationship took the approximate form (Imperial units):

$$(A_T/W^{1/3})(V/W) = 15, \qquad (V \text{ in ft}^3). \tag{5.33}$$

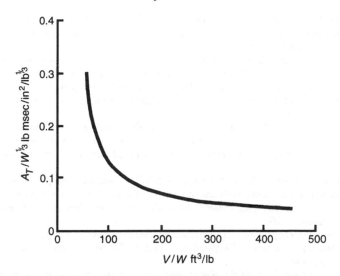

Figure 5.19 Blast in confinement (cubical enclosures) (from Christopherson, ref. 4.6).

5.6 REFERENCES

5.1 Henrych, J. (1979) *The Dynamics of Explosion and Its Use*, Elsevier, New York.

5.2 Baker, Q. A., Baker, W. E. and Spivey, K. H. (1989) Blast loading from arrays of parallel line charges, *Proc. of 4th International Symp. on Interaction of Non-nuclear Munitions with Structures*, Panama City Beach, Florida, USA.

5.3 Vaile, R. B. (1952) *Underground Explosion Tests at Dugway*, Stanford Research Institute, Stanford, California, March.

5.4 Sachs, D. C. and Swift, L. M. (1955) *Small Explosion Tests – Project MOLE*, US Air Force Special Weapon Center contract 291, Stanford Research Institute, Stanford, California, December.

5.5 Drake, J. L. and Little, C. D. (1983) Ground shock from penetrating conventional weapons, *Proc. Symp. on Interaction of Non-nuclear Munitions with Structures*, Colorado, USA.

5.6 Westine, P. S. and Friesenhahn, G. J. (1983) Free-field ground shock pressures from buried detonations in saturated and unsaturated soils, *Proc. Symp. on Interaction of Non-nuclear Munitions with Structures*, Colorado, USA.

5.7 Florence, A. L., Keough, D. D. and Mak, R. (1983) Calculational evaluation of the inclusive effects on stress gauge measurements in rock and soil, *Proc. Symp. on Interaction of Non-nuclear Munitions with Structures*, Colorado, USA.

5.8 Hicks, A. N. (1986) Explosion induced hull whipping, *Advances in Marine Structures*, ARE Dunfermline, Scotland.

5.9 Haxton, R. S. and Haywood, J. H. (1986) Linear elastic response of a ring stiffened cylinder to underwater explosion loading, *Advances in Marine Structures*, ARE Dunfermline, Scotland.

5.10 Philip, E. B. (1944) *Blast Pressure and Impulse in a Tunnel – a Note on the Latest Piezo-gauge Measurements*, UK Home Office, Research report REN 359.

5.11 Taylor, W. J. (1968) Blast wave behaviour in confined regions, *Prevention and Protection against Accidental Explosion of Munitions, Fuels and Other Hazardous Mixtures*, New York Academy of Sciences, p. 339.

5.12 Britt, J. R. (1985) Attenuation of short duration blast in entranceways and tunnels, *Proc. 2nd Symp. on the Interaction of Non-nuclear Munitions with Structures*, Panama City Beach, Florida.

5.13 Itschner, D. and Anet, B. (1977) Entry and attenuation of shock waves in tunnels, *Proc. 5th Int. Symp. on Military Applications of Blast Simulation*, Stockholm.

5.14 Gurke, G. and Scheklinski-Gluck, G. (1980) *An Investigation of Blast Wave Penetration into a Tunnel Entrance*, Report E7/80, Ernst-Mach-Institut der Fraunhofer-Gesellshaft, Freiburg, Germany.

5.15 Fashbaugh, R. H. (1991) Computer code SPIDS/Shock propagation in ducting systems utilising a PC computer, *Proc. of the 5th Int. Symp. on the Interaction of Conventional Munitions with Protective Structures*, Mannheim, Germany, April, p. 218.

5.16 Binggeli, E. and Anet, B. (1991) Experimentelle Untersuchung der Ausbreitung Iconventioneller Luftstösse in Tunnelsystemern, *Proc. of the 5th Int. Symp. on the Interaction of Conventional Munitions with Protective Structures*, Mannheim, Germany, April, p. 209.

5.17 Gregory, F. H. (1976) *Analysis of the Loading and Response of a Suppressive Shield when Subjected to an Internal Explosion*, Minutes of 17th explosive safety seminar, Denver, Colorado.

5.18 Baker, W. E. (1960) The elastic-plastic response of thin spherical shells to internal blast loading, *Journal of Applied Mechanics*, **27**(1), Series E, March.

5.19 Kingery, C. N., Schumacher, R. N. and Ewing, W. O. (1975) *Internal Pressures from Explosions in Suppressive Structures*, BRL Interim Memo Report No 403, Aberdeen Proving Ground, Maryland, USA, June.

5.20 Swisdak, M. M. and Montanaro, P. E. (1991) INBLAST – a new and revised computer code for the prediction of blast inside closed or vented structures, *Proc. of the 5th Int. Symp. on the Interaction of Conventional Munitions with Protective Structures*, Mannheim, Germany, April, p. 196.

5.21 Proctor, J. F. (1972) *Internal Blast Damage Mechanisms Computer Program*, NOL Tech Report TR 72-231, August.

5.22 Britt, J. R. et al. (1986) *BLASTINW User's Manual*, *ARA 5986-2*, Applied Research Associates Inc., Vicksburg, Miss, USA, April.

5.23 Britt, J. R. and Drake, J. L. (1987) Blast loads from internal explosions and other reflected shock waves, *Proc. of Int. Symp. on the Interaction of Conventional Munitions with Protective Structures*, Mannheim, Germany, March, p. N54.

5.24 Britt, J. R. and Drake, J. L. (1985) Propagation of short duration air blast into protective structures, *Proc. 2nd Symp. on the Interaction of Non-nuclear Munitions with Structures*, Panama City Beach, Florida, USA, April.

5.25 Kriebel, A. R. (1972) *Airblast in Tunnels and Chambers*, US DASA 1200-II Supplement 1, Defense Nuclear Agency, Washington, October.

5.26 Coulter, G. A. (1972) *Blast Loading in Existing Structures – Basement Models, BRL MR 2208*, US Army Ballistics Research Laboratory, Aberdeen Proving Ground, Maryland, August.

6

Pressure measurement and blast simulation

6.1 EXPERIMENTAL PRESSURE MEASUREMENT

The accurate measurement of dynamic loads on structures is a demanding subject, not least in that the measuring apparatus should not interfere with the load distribution it is trying to measure. The physical properties of blast waves as they strike the structure are most commonly recorded in terms of pressure, and the design and use of pressure gauges (or transducers) suitable for recording the history of blast wave pressure is an important aspect of structural loading research.

What are referred to as side-on gauges have been frequently used for blast measurement over the years. Sensing elements are mounted in housings which are contoured to offer minimum resistance or disruption of natural forces. All types of side-on gauge are unidirectional, and most use piezo-electric materials that respond to pressures on their surface. Natural materials of this type are tourmaline or quartz, and there are several synthetic materials of which lead zirconate is a good example. Piezo-electric devices have a linear response over a good pressure range, but are inclined to brittleness and will not measure statically applied loads. Considerable development of side-on blast pressure gauges, using stacks of piezo-electric discs placed in a streamlined housing, has taken place over the years in the USA, particularly at the Ballistics Research Laboratory and certain research laboratories and institutes. These developments have been described by Baker [6.1], and Figure 6.1, taken from his study, shows diagramatically the BRL side-on blast gauge (or gage). Even numbers of piezo-electric discs were interleaved with metal foil discs, and set so that there was alternate polarity. All discs of one polarity were connected via tabs to an insulated electrical lead, and discs of the opposite polarity were earthed (or grounded) to the metal of the housing. The diameter/thickness ratio of the complete gauge was greater than 10:1 to minimize the effect of the gauge profile on blast flow. The diameter of the housing was rather large, and a more

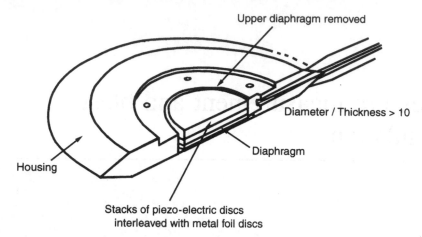

Upper diaphragm removed

Diameter / Thickness > 10

Diaphragm

Housing

Stacks of piezo-electric discs
interleaved with metal foil discs

Figure 6.1 BRL side-on piezo-electric blast gauge (from Baker, ref. 6.1).

streamlined, smaller diameter variant was developed by the US Southwest Research Institute. This was aimed at small-scale experiments.

An alternative geometry is the cylindrical side-on gauge, in which the sensing element is a hollow cylinder of lead zirconate or barium titanite that has been radially polarized. To measure side-on pressures its axis must be set in alignment with the direction of blast wave travel.

In the UK, a very common type of pressure transducer was developed and extensively used by the Ministry of Defence, in what was formerly called the Royal Armament Research and Development Establishment. This side-on gauge consisted of 12 quartz crystals clamped between two 1 inch diameter pistons. These were suspended between neoprene diaphragms and clamped around their edges to the body of the gauge. Various versions were made, to deal with high overpressures and to present a low profile. A version similar to the BRL gauge discussed above has a diameter/thickness ratio of 12:1 and is said to have omnidirectional properties, although it is not easy to see why.

Very small flush-diaphragm transducers, the use of which is not limited to blast pressure measurement, can be converted into a conventional side-on gauge form by mounting them centrally in a streamlined housing. The use of many types of piezo-electric components in side-on gauge construction has been discussed by Whiteside [6.2]. Miniature transducers can also be mounted directly on a structural component under test so that their sensing surfaces are flush with the surface of the component. For example, small gauges measuring 0.5 in in diameter have been flush-mounted in the surfaces of aerofoils subjected to blast loading. A number are described in ref. [6.1], and Figure 6.2, taken from this reference, shows the construction of a gauge developed by Baker and Ewing [6.3]. This consisted of a ceramic piezo-electric disc mounted in an epoxy resin, held within a small stainless steel housing, and the geometry of the design was chosen to minimize sensitivity to acceleration and temperature.

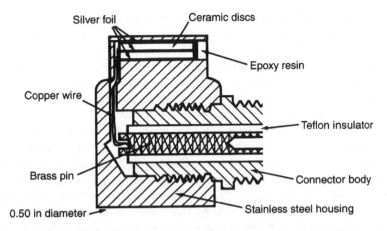

Figure 6.2 Gauge by Baker and Ewing to be flush-mounted in the surfaces of blast loaded aerofoils (from Baker and Ewing, ref. 6.3).

The measurement of reflected pressure at normal incidence to a plane surface is easier than the recording of side-on pressure, and most of the American experimental data on reflected blast waves discussed earlier was obtained from gauges designed by Hoffman and Mills [6.4]. These used piezo-electric elements mounted in a cavity in the front face of a heavy metal housing, and could measure peak pressures up to 1500 psi. An improved design by Granath and Coulter [6.5] could deal with peak pressures of 4500 psi, and according to Baker has been sold commercially.

An alternative method of pressure measurement is to sense the deflection of a ferromagnetic diaphragm under pressure by measuring the change in inductance of coils. By this means very tiny gauges can be made, having a thickness of 0.1 in thick and a diameter less than 0.2 in. The deflection of a diaphragm can also be measured by the interception of flux lines from a magnetic coil as the deflections occur. As more flux lines are intercepted the eddy current losses increase, thus changing the coil impedance. Pressure can also be sensed by supporting diaphragms on tiny tubular columns fitted with semiconductor strain gauges, and measured from the output of the gauges as the diaphragm deflects under pressure. This type of strain gauge tends to have a low sensitivity output but a high-frequency response characteristic. The diaphragms are made from many types of material, ranging from stainless steel, nickel, and stainless steel covered with Teflon or coated with flame-sprayed aluminium oxide. Baker points out that for high blast pressures near the detonation point of underground nuclear explosions the diaphragms must be protected by a heatshield baffle.

As well as the measurement of peak instantaneous pressures and the decay of pressure with time, the time of arrival of the blast wave is often required. A number of types of gauge are now available commercially which generate a large signal on arrival of the shock front but do not need to be sensitive enough to measure the pressure accurately. A very simple gauge developed by the US Ballistic Research Laboratory has a single piezo-electric disc mounted on a brass pedestal, embedded in epoxy resin and contained within a nylon housing. The signal is conveyed by a robust connector that transmits the sharp electrical charge. Of course most gauges of this type are destroyed by the explosion. Measuring devices for recording drag or dynamic pressures are also required for blast/structure interaction research, and these are usually similar in form to drag and 'total head' measuring gauges used in wind and air-flow studies. The sensing elements of total head gauges, like pitot tubes, are often simple load cells mounted within a shaped nose cone. For low or fractional Mach number flows the gauge housing is a hemispherical nose, and for higher Mach numbers the nose is conical. Drag gauges often consist of cantilever beams protruding from a base plate, which are deflected by the winds sweeping past the gauge. Measurements of bending strains in two orthogural planes enable the variation of drag with time to be recorded.

Before the coming of piezo-electric devices, most blast gauges were mechanical rather than electrical, and the Introduction to this book has a short review of the historical methods of measuring shock pressures within gun barrels by crusher gauges and the use of deforming cans to obtain data on the decay of pressure with distance in nuclear explosions. The 'deforming cans' notion gave rise in the 1960s to the development of peak-pressure gauges based on the rupture of paper or aluminium foil stretched over the open end of box-like structures which could be set face or side-on to the blast. Another simple gauge was developed in Canada at about the same time, which used the surface tension of a fluid as the principle of operation. A transient pressure high enough to break the surface film would be proportional to the surface tension and inversely proportional to the diameter of the orifice over which the film was subtended. Details of this gauge were given in 1962 by Muirhead and McMurty [6.6] who recorded a response time of 3 msec. A further idea, described by Palmer and Muirhead [6.7], employed the principle that the velocity of streamlined flow in a tube is directly proportional to the pressure causing it. A small tube was attached to an ink reservoir, the upper surface of which was open to the blast. The impact of a shock front on this surface resulted in ink squirting from the small tube on to recording paper. The horizontal distance travelled by the jet before falling on to the paper was a direct measurement of velocity.

Associated with most of the gauges described above are methods of recording and displaying the electrical signals. These are under continual development and will not be discussed in detail here. Most are linked to the Cathode Ray tube oscilloscope recording system, to magnetic tape recorders or to the galvanometer oscillograph system. Blast wave photography is also a continually developing subject, and the rotating prism high speed camera is an obvious example of instrumentation vital to explosive research. Some research laboratories employ streak cameras, which use rotating mirrors and narrow slits to focus on particular parts of the image to be studied, and the spark shadowgraph, which is useful for the photography of shock fronts. At the end of all this gauging and recording work comes the reductions of data for comparison with analysis and for codes of practice.

The measurement of shock pressures in soils has been a fruitful research field in the past thirty years. Before that time most soil pressure gauge work related to the measurement of static pressures, and many comprehensive reviews of static developments were written in the early 1960s. Examples are the paper by Selig in 1964 [6.8] and a review by Abbott *et al.* in 1967 [6.9]. By the mid-1980s experimental work on dynamic measurement had resulted in two types of pressure gauge for the recording of shock pressures in soil, the piezo-electric gauge similar in form to the air blast piezo-electric gauge, and the diaphragm deflection gauge with a diameter/thickness ratio of about 10. A new type of gauge reported at that time was the Polymer pressure gauge, described by Bur and Roth [6.10]. This is basically a piezo-electric gauge which uses poly-vinyliden fluoride (PVDF) as the transducer material, and which is also pyro-electrically active. The substance becomes active when polarized by a large electric field (2 MV/cm at room temperature). The gauge consists of two layers of PVDF sheet, each containing an active area of 10 mm diameter within an overall gauge diameter of 15 mm. Aluminium and Indium are deposited over the surfaces of the circular sheets to form electrodes, and a thermocouple is inserted between the sheets. The sheets are glued together with an epoxy and are covered with protective polycarbonate. The overall thickness of the gauge is 0.35 mm. For a sandy soil having a modulus of about 5.7×10^4 psi and a gauge modulus of about 2×10^6 psi, a modular ratio of 35 is obtained. The aspect and modular ratios satisfy the requirement for accurate stress measurement in soil, and tests show that a linear relationship exists between the input dynamic stress and the output voltage signal. For certain time ranges the gauge needs an active temperature compensation circuit because of the pyro-electrical characteristics of the material (see ref. [6.10]).

The use of polymer and other relatively recently developed materials for shock pressure gauges has been the subject of many papers in the on-going series of conferences on the Military Application of Blast Simulation, and the reader is referred to the proceedings of these conferences (MABS) for further information (e.g. ref. [6.11], Turner).

6.2 BLAST SIMULATION

Field tests to measure the dynamic loading on structures are very expensive and time consuming. The need to develop more economic test arrangements has always been a factor in studying structural response to shock, and as we pointed out earlier, much has been owed to the invention of the shock tube by Vieille [6.12] in 1899. It was the need to study blast and blast loading in the Second World War by laboratory rather than field tests that began the rapid development and use of the shock tube. Wright [6.13] in his book on the subject tells us that

Reynolds [6.14] used a shock tube to produce waves of known strength in order to calibrate the early piezo-electric gauges described in the previous section. This was necessary because static calibration was not possible. By 1945 Smith [6.15] had used photography of the shock front to study wave reflection, and then Bleakney [6.16] used an interferometer in conjunction with a shock tube to examine the diffraction of a shock wave around an obstacle.

The most simple shock tube consists of a rigid cylinder divided at a point along its length by a transverse gas-tight diaphragm. One half, the compression chamber, contains gas at a pressure well in excess of the pressure in the other half, the expansion chamber. When the diaphragm is suddenly ruptured, or caused to shatter, the shock wave heads down the expansion chamber and a rarefaction wave travels back into the compression chamber. When the shock front meets the far end of the expansion chamber it is reflected. If the end is closed the reflected wave will be a shock wave; if the end is open it will be a rarefaction wave.

Elementary shock tube theory was set down by Taub [6.17] in 1943. If it is first assumed for simplicity that the diaphragm has no effect on the system once it has ruptured, then the shock and rarefaction waves can be said to form instantly. The shock strength is p_1/p_0, where p_0 is the pressure in the expansion chamber ahead of the shock, and p_1 is the instantaneous pressure at the shock front. The maximum theoretically achievable strength y_m is given approximately by

$$y_m = \frac{2\gamma_0(\gamma_0 + 1)}{(\gamma_3 - 1)^2}\left(\frac{c_3^3}{c_0}\right),$$

(6.1)

where γ_0 is the gas constant for the expansion chamber, γ_3 is the gas constant for the compression chamber, and c_0 and c_3 are the speed of sound for the gases in the two chambers. If the gas in both chambers is air $c_0 = c_3$ and $\gamma_0 = \gamma_3 = 1.4$, so that $y_m = 42$. If the gas in the compression chamber is hydrogen, then $c_3/c_0 = 3.8$ and $\gamma_3 = \gamma_0 = 1.4$. When $p_s/p_0 = 500$ for air/air, $p_1/p_0 = 10$; when $p_s/p_0 = 500$ for hydrogen/air, $p_1/p_0 = 50$, so there is considerable gain in terms of the strength of the shock front in using hydrogen in the compression chamber. If the hydrogen is heated $c_3/c_0 > 3.8$, which increases the strength of the shock still further.

In fact the ruptured diaphragm does influence flow, so the above shock strengths cannot quite be achieved in practice. To limit the diaphragm effect as far as possible cellulose acetate sheet, which shatters into small pieces, was often used in small tubes. In larger tubes metal foil was used because the stresses in the deflecting diaphragm sheet as the compression chamber pressures were increased become too great for cellulose acetate. There is also some reduction in strength due to deceleration of the shock front as it travels down the expansion chambers. This occurs because of the effect of the boundary layer between the front and the tube walls. In well-designed tubes the influence of both of the above effects can be reduced to a level that does not greatly impair the accuracy experiments.

The properties of shock waves were measured by a number of techniques during the early years of shock tube development. A number of optical methods were used, based on the fact that the refractive index of a gas varies with its density according to the Gladstone Dale Law $\mu = 1 + K\rho$, where μ is the refractive index, K the Gladstone Dale constant, and ρ the density. The measurement of μ at high shock front speeds was often made by using short duration sparks as the light source and using shadow or Schlieren photography to examine density discontinuities. The density profile was also found by interferometry, in which light from a point source was collimated and split at a half-silvered mirror. One half then passed through the shock tube before the beams were recombined, and with monochromatic illumination interference fringes could be detected. There was a linear relationship between fringe shift and density change. High-speed rotating-mirror cameras were needed to achieve sufficient time resolution for very fast shocks.

The advent of the piezo-electric pressure gauge enabled pressures rather than densities to be measured, as long as the gauges were mounted in the shock tube in ways that eliminated the effect on them of mechanical stresses transmitted through the walls of the shock tube. This is now the most usual way of measuring shock wave properties. The instantaneous rise in temperature across a shock front can also be used to investigate density changes, so there was some development in the use of resistance thermometers and hot-wire anemometers. The latter, however, generally have too slow a response time for shock tube investigations.

We noted earlier that the strength of the shock in a tube can be increased by introducing combustion into the compression chamber. This method, which was the subject of much research in the 1950s, burns oxygen with the hydrogen to raise the temperature of the latter. The best mixture is 1 part by weight of oxygen to 8 parts by weight of hydrogen, and this leads to a sound speed (c_3) about 1.7 times that of cold hydrogen. For this application the combustion chamber must be strongly constructed, and gun barrels are frequently used. Mach numbers of about 10 are obtainable with cold hydrogen, and up to 20 with a combustion shock tube.

The heating process tends to produce non-uniform conditions in the compression chamber and attenuates the shock wave as it travels down the expansion chamber. To eliminate this problem multiple diaphragms are sometimes used, in which an intermediate chamber is introduced at a lower pressure than the compression chamber, but at a higher pressure than the expansion chamber. When the first diaphragm is ruptured the shock wave travels down the intermediate chamber until it is reflected off the second diaphragm. The reflected pressure is sufficient to rupture the second diaphragm, but the reflected wave leaves a hot, high pressure region to act as a compression chamber input which drives the shock front down the expansion chamber with a gain in strength.

Multi-diaphragms result in a noticeable increase in shock strengths. Each additional chamber is said by White to increase the shock strength by a factor of 12 or 16 according to whether the gases used have a constant (γ) of 1.67 or 1.4. These numerical increases are based on the assumption that the times of rupture of the intermediate diaphragms are long compared with the time for the reflected shock to travel an appreciable distance.

The strength of the shock is limited by the maximum speed at which gas can escape from the compression chamber of a cylindrical shock tube ($y_m = 2c_3(\gamma_3 - 1)$. This limitation can be circumvented, and the compression chamber sound speed (c_3) increased by using a compression chamber of non-uniform cross section which converges in diameter as it approaches the expansion chamber. Alpher and White [6.18] suggested that increases of up to 10% in Mach number can be achieved by this method. The strengthening effects of convergence are also attainable by tapered reductions in the diameter of the expansion chamber.

The final interesting development of the 1950s was the replacement of the 'piston' of gas in the compression chamber by electromagnetic forces. An electromagnetic piston, having a velocity limited only by the speed of light, was established by the introduction of two parallel electrodes, across which a high voltage was suddenly applied. In argon gas at an ambient pressure of 1 cm of mercury a potential gradient of 100 volts/cm will generate Mach 20 shock fronts, and at very, very short durations it is even possible to achieve Mach 200.

In using the shock tube to apply instantaneous loading to structures, there is clearly a problem of scale as well as duration. This was very apparent in the development of methods of simulating the effect of a nuclear explosion on civil and military equipment. A very large diameter tube is required to dynamically load a complete tank hull or ship superstructure, and the duration of the positive period of the explosion is required to be much greater than that from a conventional charge. These difficulties have led to a considerable research interest in blast simulation methods since the 1960s, including the use of vertical shock tubes.

By the mid-1980s several large air-blast simulators had been built in various countries as part of the well-financed defence requirement to measure the blast loads from nuclear explosions on full-sized military equipment such as tanks, small aircraft and helicopters. Most of these were shock tubes in which cross-sectional areas ranged from 20 to 180 m^2, and total lengths were up to 100 m, or even 200 m. One of the first large air blast simulators to be built was at the Atomic Weapons Research Establishment (as it was then called), at Foulness Island, England, and has been described in a number of papers by Leys [6.19]. It was completed during the 1970s, and the test section was 20 m^2 in area. It was originally built with an explosive driver developed from two naval guns which fired into a 1.82 m diameter steel tunnel 36 m long. This was linked by an asymmetric cone to a 2.43 m diameter section 44 m long, and a further expansion cone connected this section to a 4.86 m diameter section 66 m long. This was followed by a semi-circular 10.7 m diameter extension. The development and calibration of the simulator has been reviewed by Tate [6.20] who noted that the guns were removed and the blast wave was eventually generated in the 1.82 m tube by the detonation of up to 100 m of detonating cord (59 g of PETN/m) wound vertically on an expanded polystyrene former. The use of explosives rather than compressed air meant that the diaphragms of the early small shock tubes were no longer required. The blast wave shape was found to be very similar to the 'Friedlander' shape discussed earlier (see Figure 1.1), although for economic reasons the larger tubular sections were shorter in length than would be normally adopted in shock tube work. The complete facility is shown in Figure 6.3, taken from ref. [6.20].

We have already noted that a rarefaction wave is produced at the open end of a shock tube, and considerable research has taken place to support the development of suitable Rarefaction Wave Eliminators (RWE). In the Foulness tube the eliminators take the form of metal grids that give partial blockages at the open end of the final tunnel and at the end of the 4.86 m diameter section. The blockage due to the grid at any section is about 40% of the total cross-sectional area, and Tate points out that a restriction less than this would allow too strong a rarefaction wave to propagate back into the preceding tube; a greater restriction would limit the peak pressure attainable in the final semi-circular test section. The ratio of the peak instantaneous pressures in the sections of the tube are, to a first approximation, inversely proportional to the ratio of the cross-sectional areas. At the time of Tate's paper the peak pressure in the final semi-circular tube was limited to less than 55 psi.

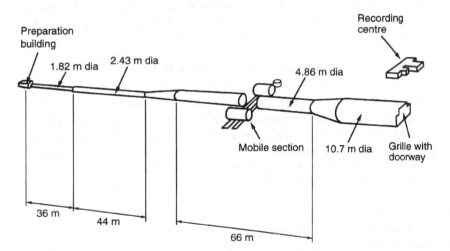

Figure 6.3 Explosive-driven blast-wave simulator, Foulness, UK (from Tate, ref. 6.20).

The development of the Foulness simulator demonstrates the various aspects of detail design, equipment and instrumentation that must be taken into account in the construction and operation of large shock tubes. For example, there was considerable experimentation and discussion before the method used to drive the simulator was finally evolved, as reported by Clare [6.21]. The 'smoothing' of high-frequency perturbations that affect the shock wave/duration profile was important, and research was devoted to the use of an aqueous foam plug in the 2.4 m diameter section to attenuate the 'spikes' in the wave form. The 'side-on' pressure gauges to measure static pressures in the main test sections have been described by Leys [6.19] and dynamic differential pressure gauges were developed and described by Ethridge *et al.* [6.22]. The Foulness facility simulates an equivalent free-air high explosive yield of 1.5 kt of TNT, which has a peak overpressure of up to 50 psi and a positive phase duration of 350 msec. The simulator reaches the required pressure, but the positive phase durations are somewhat less than 350 msec. Depending on the use of smoothing and rarefaction eliminators, the durations are between 150 and 300 msec.

The compressed air system, rather than the explosive methods of driving shock tubes, has been retained by the French and German authorities in their blast simulator development. In the 1970s the French installed an air-driven simulator at the Centre d'Etudes de Gramat, which consisted of a single large semi-circular tube, 100 m long, 70 m^2 in cross-sectional area, driven by a bank of single shock tubes of less than 1m in diameter. The shock fronts emerging from the individual driver tubes are mixed and diffracted, and this causes a very spiky pressure/duration relationship. This must be smoothed as far as possible in a zone of the tube between the driver tubes and the test section as indicated in Figure 6.4.

At the far end of the tube a reflection wave eliminator in the form of a grid is used. The conception and performance of this simulator has been described by Gratias and Monzac [6.23], and research into methods of damping or eliminating the perturbations in the pressure/time curves by the use of perforated tunnels and extension cones has been reported by Hoffman [6.24].

The driver tubes were originally separated from the driven zone by metal diaphragms, which were broken instantaneously by pyrotechnic devices. The safety rules associated with the installation of these devices, and the possibility that pieces of the diaphragms might break off and become projectiles down the tube, led to research in diaphragm replacement, reported in 1985 by Gratias [6.25]. It was proposed that a valve system be used with a very short opening time of about 20 msec. A parallel study on fast acting valve systems was made at the same time by Osofsky and Mason [6.26], as part of the feasibility study for a US military large shock tube test facility. They concluded that heated driver gas was required to produce good wave forms if the diaphragm system was used. In the diaphragm system unheated driver gas produced dynamic pressure discontinuities, particularly at high overpressures (735 psi). If, however, fast acting valves were used in place of diaphragms, it was found

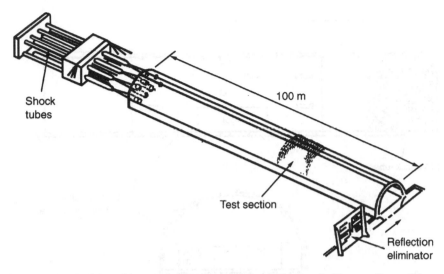

Figure 6.4 Air-driven blast wave simulator, Centre d'études de Gramat, France (from Gratias and Monzac, ref. 6.20).

possible to use unheated driver gas to achieve smooth wave forms at high pressures, and to control positive phase duration. It was thought that a rotary drum fast-acting valve was a practical way to mechanize the wave-shaping technique. The schematic driver arrangement for the US facility is shown in Figure 6.5.

The German blast simulator was formed by driving a tunnel into the rock of the Reiter Alpe range of hard chalkstone. The tunnel is 106 m long and semi-circular in section with a total cross-sectional area of 76 m^2 and the general arrangement is shown in Figure 6.6, taken from a paper by Ackermann and Klubert [6.27]. The blast wave generator consists of 144 shock tubes, combined into a battery. The compressed air bottles are clamped horizontally into a frame, and in contact with a wall so that the thrust resulting from the bursting of the

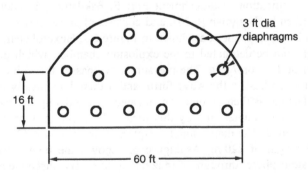

Figure 6.5 Schematic driver arrangement in large blast/thermal simulator (US proposal) (from Osofsky and Mason, ref. 6.26).

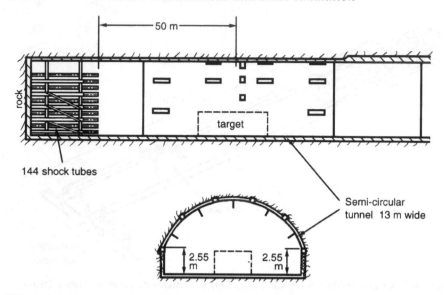

Figure 6.6 Shock tube driven blast simulator, Reiter Alpe, Germany (from Ackerman and Klubert, ref. 6.27).

diaphragms can be absorbed by the rock. Each tube has an interior volume of 0.4 m³, and is safe to 200 bar. The diaphragms are made from hardboard and fired by plastic explosive. The structure testing area is about 50 m from the battery of driver tubes, but experiments showed that at this point the side-on pressure/time relationship is not ideal. A sharp instantaneous pressure rise is followed by exponential decay, but then there is another sudden pressure increase lasting for 50 msec. Because of this pressure increase the reflections from the walls overlap the blast wave and subsequent exponential decay suddenly ends after a time of 300 msec. During the early stages of use a 0.8 bar shot resulted in damage to the walls and ceilings of local village houses.

Another blast simulator embedded in rock was constructed in the mid-1960s at the Swedish fortifications establishment Fort F, Eskilstuna, Sweden. It has been fully described by Bergman [6.28], and consists of an explosion chamber, test chamber and exhaust tunnels, as shown in Figure 6.7. Hexotol charges up to 100 kg in weight can be detonated in the explosion chamber, which give peak gas pressures of up to 6 or 7 kPa. The pressure wave passes through a diffuser which reduces the 'spikes' in the wave form, and it then dynamically loads the test specimen. The gas is then gradually evacuated by regulating the size of the entrance to the exhaust tunnel, and by this means the duration of the pressure pulse can be controlled. The main tunnel is square in cross section (2.1 m sides) and has a total length of 220 m. As Figure 6.7 shows, the tunnel eventually discharges to atmosphere through a muzzle system. By disconnecting the exhaust loop, removing the diffuser and emptying the test section of equipment, a straight shock tube is obtained. The gas flow from the entrance can then load

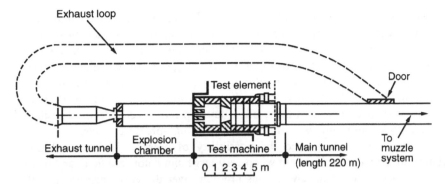

Figure 6.7 Blast simulator embedded in rock, Eskilstuna, Sweden (from Bergman, ref. 6.28).

large military objects placed on level ground immediately outside the end of the tunnel, where the stagnation pressure can reach 300 kPa with a duration of about 0.5 sec. However, the side-on pressure has only about 0.01 sec positive duration.

Instead of detonating a single charge in the explosion chamber, a line charge can be detonated in the main tunnel. The charge is normally constructed of cordtex filaments that are either hung on posts or wound on cylindrical formers of plastic or steel. Two types of cordtex were used in a test programme reported by Bryntse [6.29], Nobelcord at 10 g PETN/m and E-cord at 5 g PETN/m. The muzzle at the entrance, which cancels the rarefaction wave, is made as a hollow concrete block with a number of bricks that can be added or removed to adjust the size of the opening. In the tests the open area of the muzzle was varied between 0.25 and 2.25 m^2.

There was an approximately linear relationship between charge length (up to a length of 100 metres) and positive duration (up to 300 msec), but a Friedlander-type pressure/duration relationship could only be obtained for higher pressure levels. The lower pressure limit was about 8 kPa reflected pressure, below which a line of cordtex tends to give a square-formed pressure pulse; this problem can be alleviated somewhat by helically winding the cordtex on a 0.3 m diameter cylindrical polystyrene former. The test results were analysed by Bryntse, who compared the actions following the detonation of a concentrated charge and a distributed (or line) charge in a tube with smooth walls. For concentrated charges two exponential waves will propagate along the tube, after some complex reflection processes, with a gradual change in the pressure/duration relationship. For a distributed charge the reflection interaction with the tube walls will occur over a much greater length of the tunnel, and the outward moving pressure pulses will be of long duration and of approximately square wave shape. The wavelength of each pulse will be equal to the charge length.

The tests indicated that the detonative combustion process for line charges was somewhat incomplete, and was strongly influenced by imperfections in the cordtex production process. Although the PETN content per unit length was thought to vary by no more than ±10%, small variations in packing density were found to cause large variations in released energy on detonation. Because of this the relationship between impulse and charge weight contained a much wider experimental scatter. The relationship was also thought to be influenced by the changes in shock wave shape, because ripples appear in the original rectangular-shaped pulse, and these often develop into several peaks. The final simple exponential profile was only achieved at high-pressure levels.

We have already mentioned the possibility of noise and vibrational disturbance to the local environment when very large simulators are fired, and there have been a number of reviews of this problem over the years, beginning with an investigation by James [6.30] in 1965 and continuing more recently with studies by Cadet [6.31] in France and Reed [6.32] in the USA. James concluded that incident over pressure (p_i) in psi at a distance d feet from the mouth of a shock tube, and at an angle θ to the tube axis, was given approximately by the relationship

$$\log p_i = -0.00787\theta - 1.257 \log d + 0.987 \log p_w + 0.536, \qquad (6.2)$$

where p_w is the internal overpressure in the shock tube. To a close approximation this indicates a linear relationship to internal pressure and that decay with distance is close to the theoretical prediction. A more general expression derived from the data of James, but including the effect of the dimensions of the end of the tube, is given by Reed in SI units of pascals and metres as

$$p_i = 0.186 \, p_w (A^{1/2}/R)^{1.1} \, e^{1.326 \cos \theta}, \qquad (6.3)$$

where A is the cross-sectional area of the tube in m^2 and R the distance in metres. This assumes an overpressure-distance decay exponent of -1.1. Tests showed that for a large shock tube (5.8 m diameter) at the Sandia laboratory in the USA (driven by 500 kg of primacord), and for a tunnel shock tube simulator (cross-sectional area of the tunnel portal of 28.6 m^2) at the Nevada Test Site of the US Defense Nuclear Agency, the distant air blast propagations could be reasonably predicted by this equation (6.3). Local characteristics, such as woods and hills, affect these propagations, but the equation gives a useful indication of overpressures at distances up to 5 or 10 kilometres from the simulators. The assessment of environmental effects was particularly important at the Sandia site, because the driver end of the shock tube was open to the east, and pointed at the Solar Thermal Test Facility. This incorporated 3000 glass mirrors and was about 1.8 kilometres away! The cross-fertilization of scientific research does not normally involve years of bad luck through cracked mirrors!

So far we have reviewed the experimental simulation of air blast, but in assessing the survivability of equipment in nuclear explosions the possible effects of heat cannot be ignored. Several research projects in the 1980s were aimed at the provision of a thermal radiation source in existing or projected large air blast simulators. The source was usually a hot-burning plume or flame from a mixed liquid oxygen and powdered aluminium jet, described for example by Haasz, Gottlieb and Reid [6.33], Pearson, Opalka and Hisley [6.34] and Bogartz [6.35] in the ninth International Symposium on the Military Applications of Blast Simulation. The hot burning torches are usually set in line across the test section of the simulator, so that a flat radiating flame surface exists just ahead of the target. This flame burns with a high thermal output of typically 250 J/cm^2/sec for 5 seconds, and because of the high volume of hot reaction products the tube must be fitted with a rapid venting system. Each kilogram of burning powdered aluminium produces about four cubic metres of gaseous and solid-particulate cloud, and if this is not removed it will occupy too great a volume of the simulator channel with consequent distortion of the blast wave amplitude and shape. Bogartz [6.35] suggests an air mover based on the venturi principle, four of which are installed in the roof of the tube. Each mover extracts 5 m^3/sec, and is helped in this action by the installation of air curtains issuing from floor slots. These curtains confine and convect the reaction products upwards, and in addition help to cool the hot reaction products.

So far we have discussed simulators for nuclear or large HE explosions, where the action of the explosion is to produce a shock wave travelling at a speed greater than that of sound. However, we know from Chapter 1 that deflagration, in which shock waves do not form, is typical of some unconfined fuel-air explosions and also dust explosions. An account of a simulator for deflagrating explosions has been given by Hoffman and Behrens [6.36] where it is pointed out that the deflagration pressure-time relationship can be approximated to a triangular positive pulse with a linear increase from zero followed by a sudden decay to a negative pressure peak; then a suction phase having the same duration as the positive phase, during which the pressure returns linearly to zero. To achieve this pressure-time history experiments were made with slow-burning gunpowder and a compressed-air-driven piston in a tube. Eventually, however, the solution took the form of a two-chamber simulator consisting of a single tube 2.4 m in diameter separated into two equal halves by a wall with a square opening at the centre, as shown in Figure 6.8. The specimen plate to be loaded is clamped across this opening on a support frame. In the end plate of Chamber 1 is a small circular opening closed by a membrane, and in the end plate of Chamber 2 is a larger opening, also closed by a membrane.

Both chambers are then charged with compressed air to a pressure of twice the peak pressure required in the test. At time $t = 0$ the small membrane at the end of Chamber 1 is destroyed instantaneously by a small high explosive charge. Air escapes and the pressure in Chamber 1 decreases. As the pressure in Chamber 1 drops to half its value the membrane at the end of Chamber 2 is destroyed. The pressure in Chamber 2 drops very rapidly through the large opening. As the reducing pressure in Chamber 2 'overtakes' that in Chamber 1, the resulting pressure in the test specimen is again equal to zero. It then reverses in direction, and reaches a maximum negative pressure when Chamber 2 is exhausted. An advantage of this system is that for calibration purposes the specimen under test can be loaded statically by using Chamber 2 only and filling it slowly with compressed air. Further, Chamber 2 acting alone can be used to simulate a detonation explosion by suddenly filling it with compressed air. An instantaneous pressure rise cannot be achieved of course, but a rise of 3 to 5 msec is acceptable if the test specimen has a long duration natural period of oscillation. The rate of decay of pressure after the peak has been reached can be controlled by the size of the opening in Chamber 2. A small opening gives a pressure-time curve shape not unlike that from a nuclear weapon, and a large opening gives a relationship similar to that of a conventional explosion.

6.3 SIMULATION OF LOADS ON UNDERGROUND STRUCTURES

The application of shock loads to the surface of soils under which protective military structures have been buried became important in the 1960s, when the moratorium on full-scale nuclear weapon tests meant that other means of accumulating experimental data had to be found. The problem was attacked in several ways and on several scales. Some of the large air-blast simulators

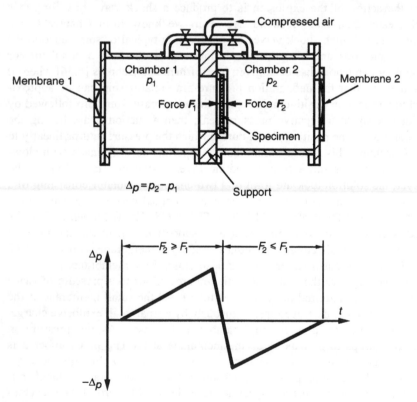

Figure 6.8 Construction and method of operation of a deflagration simulator (from Hoffman and Behrens, ref. 6.36).

described in section 6.2 were fitted with test sections in which the travelling shock front passed over a mass of soil. Various model structural forms could be built within the mass, and subjected to shock loading via the soil cover. A typical example of this arrangement was the blast simulator at Foulness Island in the UK, where a short length of the 8 ft diameter section ran over a tank of soil. At this point the lower wall of the simulator tunnel was removed so that the shock front ran transversely over the soil surface. A cross section of the tube at the test zone was given in an earlier book by the author, Bulson [6.37], and is reproduced here in Figure 6.9. Similar arrangements were made in horizontal blast simulators in Europe and the USA.

A second type of test apparatus, for smaller-scale work, was to use a vertically mounted shock tube, the open lower end of which was set immediately above a mass of soil. The initiation of a shock wave resulted in the application of an instantaneous pressure vertically over the surface of the soil. Simple model structures could be buried in the soil, and the loading on them inferred from strain and acceleration records. A vertical shock tube facility of this type was also built at Foulness, and has been described in a report by Clare [6.38]. The author used this shock tube to apply loads to buried thin-walled cylinders, and the experiments are described in reference [6.37]. The shock tube was 40 feet long, 2 feet in diameter, and was fired by a coiled charge of cordtex at the top. Clare showed that the blast pressure-duration relationship closely followed the Friedlander form. The soil tank introduced under the lower opening of the tube was 5 feet square, and the test structures were set in various soils so that they could be viewed through tunnels. High speed cine-film was taken through the tunnels of each collapse sequence. The firing procedure was first to

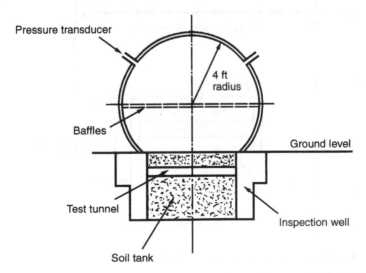

Figure 6.9 Cross section of shock tube at the test zone (Foulness, UK) (UK Atomic Weapons Research Establishment, 1980) (from Bulson, ref. 6.37).

set the camera working by electrically operated remote control from a blast proof building a short distance from the rig; a few milliseconds later the camera operated an electrical charge-firing circuit. The moment of firing was recorded by a flash on the film from a photoflood bulb operated by the firing circuit. The camera speed was 3000 frames per second, with a running time of about 3 sec, including acceleration and deceleration. The time interval from firing to complete decay of the blast pressure was about 20 msec. Blast pressures in the tube were measured by piezo-electric gauges.

A dimensionally similar vertical shock tube was constructed at the US Air Force Weapons Laboratory, Kirkland, New Mexico, in the 1960s, and has been described by Abbott [6.9]. Detonation was by means of 4.57 lb of primacord, and the shock pressure was applied to a cylindrical soil bin 48 in high, 22 in diameter.

Another type of vertically set apparatus for simulating blast loads was built at the US Waterways Experiment Station at Vicksburg, Miss. USA, and has been described briefly by Flathau, Dawsey and Denton [6.39]. It consisted of a vertical cylinder, containing a piston that could be hydraulically actuated to provide short duration concentrated loads over a maximum stroke of 4 in. Tests showed that loads in excess of 200 000 lb could be applied with a minimum rise time of 1.3 msec with a movement of about 0.25 in of the loading ram. A second version of this vertical loading system, operated in a similar way, was capable of applying loads of 500 000 lb over a time of 80 msec.

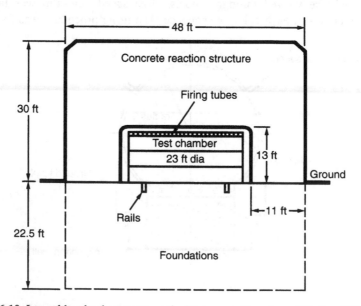

Figure 6.10 Large blast load generator at the Waterways Experiment Station, USA (from Flathau *et al.*, ref. 6.39).

The US Waterways Experiment Station was also the site of one of the most ambitious soil loading facilities ever constructed, the large Blast Loading Generator, described in ref. [6.39]. The facility, commissioned in the mid-1960s, took the form of a central firing station and test chambers, as shown in Figure 6.10, and was designed primarily to test underground protective structures subjected to pressures typical of kiloton and megaton nuclear devices. The structure consisted of a massive post-tensioned prestressed concrete reaction portal, which was basically a 48 ft × 28 ft × 28.4 ft deep slab with a 13 ft deep tunnel cut through it. Into this tunnel fitted the circular test chamber, about 23 ft diameter, containing the soil medium and the structure under investigation. The chamber was formed by stacking three steel rings, adding a further ring containing firing tubes, and then adding a top lid or 'bonnet'. High explosive was placed in the tubes and a baffle system was used to ensure that the soil surface was uniformly loaded on detonation. The vertical reaction from the exploding charges was taken on the concrete reaction structure. It was stated that peak pressures from 5 to 400 psi with rise times of between 2 and 4 msec, and positive phase duration times of several seconds could be reproduced in this generator.

For smaller-scale work the US Waterways Experiment Station also installed a Small Blast Load Generator, which was designed and constructed by Boynton Associates [6.40] and is shown diagrammatically in Figure 6.11. The steel container was a 9/16 in thick cylindrical shell with a dome-shaped top. The walls were made from stacked rings, all about 4 ft diameter but having different heights. Combinations of rings were selected to give a range of chamber heights. The soil mass consisted of a column of sand 4 ft diameter, extending 9 ft below floor level. For explosive loading primacord was detonated in two firing tubes

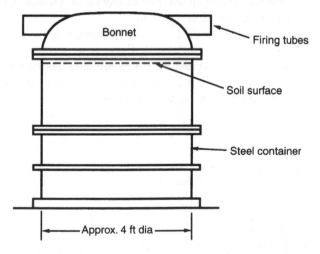

Figure 6.11 External view of small blast load generator, Waterways Experiment Station, USA (from Boynton Associates, ref. 6.40).

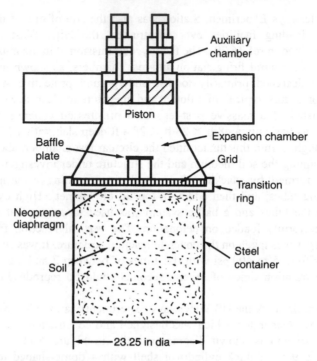

Figure 6.12 Blast load simulator at the University of Illinois, USA (from Egger, 1957, ref. 6.41).

incorporated in the 'bonnet', giving an overpressure/duration relationship with a rise time of 0.30 msec, as indicated in ref. [6.42].

The final test apparatus that we should review here was designed at the end of the 1950s at the University of Illinois, USA, by Egger [6.41] and later modified by Sinnamon *et al.* [6.42]. It is operated by a high-pressure gas system, which gives a peak surface pressure of 500 psi uniformly over the soil surface, and is shown diagrammatically in Figure 6.12, taken from ref. [6.42].

6.4 REFERENCES

6.1 Baker, W. E. (1973) *Explosions in Air*, University of Texas Press, USA.
6.2 Whiteside, T. (1967) *Instrument Development Section Notes*, Atomic Weapons Research Establishment, Foulness, UK.
6.3 Baker, W. E. and Ewing, W. O. (1961) *Miniature Piezo-electric Gauges for Measuring Transient Pressures on Airfoils*, BRL Memo Report No 1329, Aberdeen Proving Ground, Maryland, USA.
6.4 Hoffman, A. J. and Mills, S. N. (1956) *Air Blast Measurements about Explosive Charges at Side-on and Normal Incidence*, BRL report No 988, Aberdeen Proving Ground, Maryland, USA.

6.5 Granath, B. A. and Coulter, G. A. (1962) *BRL Stroke Tube Piezo-electric Blast Gauges,* BRL Tech Note No. 1478, Aberdeen Proving Ground, Maryland, USA.

6.6 Muirhead, J. C. and McMurty, W. M. (1962) Surface tension gauges for the measurement of low transient pressures, *Rev. Sci. Instr.,* **33**(12).

6.7 Palmer, W. O. and Muirhead, J. C. (1969) A squirt gauge for peak blast pressure indication, *Rev. Sci. Instr.,* **40**(12).

6.8 Selig, E. T. (1964) A review of stress and strain measurement in soil, *Proc. Symp. on Soil-structure Interaction,* University of Arizona, Tucson, USA, September.

6.9 Abbott, P. A., Simmons, K. B., Reiff, C. M. and Mitchell, S. (1967) Recent soil stress gauge research, *Proc. Int. Symp. on Wave Propagation and Dynamic Properties of Earth Materials,* University of New Mexico, Albuquerque, August.

6.10 Bur, A. J. and Roth, S. C. (1985) A polymer pressure gauge for dynamic pressure measurements, *Proc. 2nd Symp. on the Interaction of Non-nuclear Munitions with Structures,* Panama City Beach, Florida, April.

6.11 Turner, R. G. (1985) Transducers old and new, *Proc. 9th Int. Symp. on Military Applications of Blast Simulation,* St Edmund Hall, Oxford, UK, September.

6.12 Vieille, P. (1899) *Comptes Rendus,* 129, 1228.

6.13 Wright, J. K. (1961) *Shock Tubes,* Methuen (London).

6.14 Reynolds, G. T. (1943) *A Preliminary Study of Plane Waves Formed by Bursting Diaphragms in a Tube,* OSRD Report 1519.

6.15 Smith, L. G. (1945) *Photographic Investigation of the Reflection of Plane Shocks in Air,* OSRD Report 6271.

6.16 Bleakney, W., White, D. R. and Griffith, W. C. (1950) Measurements of diffractions of shock waves and resultant loading of structures, *Journal of Applied Mechanics,* **17**, 493.

6.17 Taub, A. H. quoted by Reynolds, G. T. in ref. [6.14].

6.18 Alpher, R. A. and White, D. R. (1958) Flow in shock tubes with area change at the diaphragm section, *Journal of Fluid Mechanics,* **3**(5), February.

6.19 Leys, I. C. (1983) AWRE Foulness nuclear air blast simulator: construction and calibration of the enlarged facility, *Proc. 8th Int. Symp. on Military Applications of Blast Simulation,* Spiez, Switzerland.

6.20 Tate, J. (1985) AWRE Foulness nuclear air blast simulator: development and calibration of the facility, *Proc. 9th Int. Symp. on Military Applications of Blast Simulation,* St Edmund Hall, Oxford, UK, September.

6.21 Clare, P. M. (1979) Methods used to drive the AWRE Foulness nuclear blast simulator, *Proc. 6th Int. Symp. on Military Applications of Blast Simulation,* Cahors, France.

6.22 Ethridge, N. H., Teel, G. D. and Reisler, R. E. (1983) A differential pressure gauge for measurement of dynamic pressure in blast waves, *Proc. 8th Int. Symp. on Military Applications of Blast Simulation,* Spiez, Switzerland.

6.23 Gratias, S. and Monzac, J. B. G. (1981) Le simulateur de Souffle a Grand Gabarit du Centre d'Etudes de Gramat, Conception, Etudes, Performances, *Proc. 7th Int. Symp. on Military Applications of Blast Simulation,* Suffield, Canada.

6.24 Hoffman, G. (1985) Simulation of real weapon-effects in multiple-driver shock tubes, *Proc. 2nd Symp. on the Interaction of Non-nuclear Munitions with Structures,* Panama City Beach, USA, April.

6.25 Gratias, S. (1985) Test of a fast opening valve for the 2.4, shock tube at the CEG, *Proc. 9th Int. Symp. on Military Applications of Blast Simulation,* St Edmund Hall, Oxford, UK, September.

6.26 Osofsky, I. B., Mason, G. *et al.* (1985) Wave shaping by valve motion, *Proc. 9th Int. Symp. on Military Applications of Blast Simulation,* St Edmund Hall, Oxford, UK, September.

6.27 Ackermann, J. and Klubert, L. (1985) The new large blast simulator of Reiteralpe Proving Ground, *Proc. 9th Int. Symp. on Military Applications of Blast Simulation*, St Edmund Hall, Oxford, UK, September.

6.28 Bergman, S. G. A. (1967) The RSFA underground shock tube facility for testing fortification equipment, *Proc. Conf. on Military Applications of Blast Simulators*, Defence Research Establishment Suffield, Canada, Vol. 1.

6.29 Bryntse, A. (1985) Detonating fuse in the FortF shock tube III to obtain air shock waves with low amplitude and long duration, *Proc. 9th Int. Symp. on Military Applications of Blast Simulation*, St Edmund Hall, Oxford, UK, September.

6.30 James, D. J. (1965) *An Investigation of the Pressure Wave Propagated from the Open End of a 30 × 18 in Shock Tube*, AWRE Report No 0-60/65, AWRE Aldermaston, UK, September.

6.31 Cadet, A. (1983) Investigation of the pressure wave and noise measurements outside the Large Blast Simulator, Centre d'Etudes de Gramat, France, *Proc. 8th Int. Symp. on Military Applications of Blast Simulation*, Spiez, Switzerland, June.

6.32 Reed, J. W. (1980) Project PROPA-GATOR – Intermediate range explosion airblast propagation measurements, *19th Explosives Safety Seminar*, Vol. II, DOD Explosives Safety Board, Washington, DC, USA, September.

6.33 Haasz, A. A., Gottlieb, J. J. and Reid, L. D. (1985) Air-curtain system for blast wave simulators to remove combustion products from thermal radiation sources, *Proc. 9th Int. Symp. on Military Applications of Blast Simulation*, St Edmund Hall, Oxford, UK, September.

6.34 Pearson, R., Opalka, K. and Hisley, D. (1985) Design studies of drivers for the US Large Blast/Thermal Simulator; from same source as ref. [6.33].

6.35 Borgartz, B. O. (1985) AWRE Foulness combined blast and thermal radiation simulator, installation of the TRS in the blast tunnel; from same source as ref. [6.33].

6.36 Hoffman, G. and Behrens, K. (1983) Simulation of pressure waves and their effects on loaded objects; Part 1: outlining the problem, description of the simulation device, *Proc. Symp. on the Interaction of Non-nuclear Munitions with Structures*, US Air Force Academy, Colorado, USA, May.

6.37 Bulson, P. S. (1985) *Buried Structures – Static and Dynamic Strength*, Chapman & Hall, London.

6.38 Clare, R. (1966) *A Face-on Vertical Blast Loading Simulator for Structural Response Studies*, Atomic Weapons Research Establishment, UK, Report E1/66.

6.39 Flathau, W. J., Dawsey, J. V. and Denton, D. R. (1968) Blast load generator facilities and investigations of dynamically loaded concrete slabs, *Prevention and Protection against Accidental Explosion of Munitions etc., Annals of New York Academy of Sciences*, **152**, Art. 1.

6.40 Boynton Associates (1960) *Operation Manual for 250 psi 4 foot Diameter Dynamic Load Generator*, Boynton Associates, La Canada, California, USA.

6.41 Egger, W. (1957) *60 kip Capacity, Slow or Rapid Loading Apparatus*, Dept of Civil Engineering, Univ of Illinois, USA, Report SRS 158.

6.42 Sinnamon, G. K. and Newmark, N. M. (1961) *Facilities for Dynamic Testing of Soils*, Dept of Civil Engineering, Univ. of Illinois, USA, Report SRS 244.

7

Penetration and fragmentation

7.1 INTRODUCTION

The analysis of dynamic impact is an important feature in the study of explosive effects on structures, partly because military or terrorist activities frequently involve the transportation of explosive charges by high-speed projectiles, and partly because the explosion of the charge is often accompanied by the high-speed distribution of fragments of disintegrating containers. Sometimes the disintegration of one part of a bombed structure sends fragments of that structure over a larger area.

An explosive warhead can be aimed at a structure by a variety of means: aerial bombs, rockets, artillery, missiles. The degree to which these warheads penetrate a surface structure, or the earth adjacent to a buried structure before exploding, has a large effect on the eventual damage. We therefore need to survey the state of research in the fields of bomb or missile penetration into a range of materials, from soil to concrete, and from steel sheet to sophisticated armour plating. We must take account of the possible deformation and crumpling of the charge casing during the penetrative process, because this will affect the penetration depth.

The fragmenting casing usually projects small pieces of virtually undeformable metal at high velocities. When these strike solid structures such as protective concrete bunkers considerable surface damage may occur. When these solid fragments strike relatively light structures, like aircraft wings or the cellular walls of naval craft, the fragments remain in good shape, but the sheet metal is penetrated and deformed. Certain military weapons like high-velocity cannons can spray large areas of plated structures with projectiles that produce a pepper-box type of damage to large areas of unsupported sheeting. If the holes are irregular in shape, and occur in brittle material, fast-growing cracks can be formed. These can limit the residual life of the structure under cyclic loading.

The original scientific investigations of dynamic penetration were linked to the development of military weapons in the eighteenth and nineteenth centuries, and this section will therefore include a historical reference to the research work of that era. One of the earliest penetration theories [7.1] was due to J. V. Poncelet (1788–1867), mentioned earlier in the Introduction, who was a French military engineer and professor of engineering mechanics at the military school in Metz. He had a perceptive view of the physics of the penetrative process, in which he combined the resistance of a medium to the static penetration by a body with a resistance linked to the speed at which the body was moving. He considered that the dynamic resistance in a material such as soil or sand should be analogous to resistance to fluid dynamic flow.

He therefore assumed that the force resisting penetration would be given by the equation

$$F = -A(a + bV^2), \tag{7.1}$$

where

F = resisting force
A = cross sectional area of the projectile
V = impact velocity
a, b were constants to be determined by experiment.

If the mass of the projectile $= m$, the equation of motion is

$$m\frac{dV}{dt} = -A(a + bV^2), \tag{7.2}$$

and by integration we get the relationship

$$p = \frac{m}{2ab} \log_e\left(1 + \frac{bV^2}{a}\right), \tag{7.3}$$

where p = vertical penetration.

This assumes that the material has sufficient depth in the direction of penetration to bring the projectile to rest before it 'breaks out' of the far side. If this is not so, adjustments have to be made to the analysis, as we shall see later. The difference between penetration into soil or concrete, or other frangible materials, and into strong, ductile materials is very great, and is influenced by the average pressures resisting penetration. These can be 15 times greater in steel armour, for example, than for concrete. For all materials it is the kinetic energy of an undeforming projectile that crushes and distorts the target.

The contribution of Poncelet to many problems in materials science has been well documented in the book *History of Strength of Materials* (ref. [7.2]), and it is from this that we learn that he started from poor beginnings before winning scholarships that eventually took him to the Ecole Polytechnique. He joined Napoleon's army in 1812 and was taken prisoner during the retreat from Moscow. His ideas about projective geometry were developed during his confinement for two years in Saratov. After his subsequent career at Metz he became commandant of the Ecole Polytechnique between 1848 and 1850.

His very early investigations of the effects of dynamic loads were of great fundamental value. He appreciated the importance of ductility in the absorption of kinetic energy and he was the first to demonstrate that a suddenly applied load produces double the stress of the same load applied gradually. He also demonstrated the danger of vibrational resonance and of fatigue. There is little doubt that he was a genius, and his capacity for uncovering the roots of structural and mechanical behaviour have been of great value in many fields.

7.2 PENETRATION INTO SOIL, STONE AND ROCK

Penetration into soil of falling bombs and artillery shells was a particularly important research area supporting the development of military weapons during the nineteenth and twentieth centuries, and many thousands of experiments at large scales were carried out. For vertical penetration into well-compacted soil the results confirmed (but only to an accuracy of ±20%) that Poncelet's V^2 term was suitable when impact velocity exceeds about 60 m/sec. At the beginning of the nineteenth century many empirical expressions for the constants a and b were developed, but later Petry [7.3] in 1910 proposed a modified formula,

$$p = \frac{W_p}{A} \cdot K \log_{10}[1 + V^2/215\,000], \tag{7.4}$$

where

p = depth of vertical penetration in feet
W_p = total projectile weight in lb
A = cross-sectional area of the projectile in in^2
K = constant depending on target material
V = striking velocity in feet/sec.

The constant, K, was found to depend on V, as well as on the soil properties. In the units of Eq. (7.4), the following approximate relationships were recommended for a broad range of typical soils:

Sand: $K = 154 - 0.07V$
Sandy loam: $K = 190 - 0.09V$
Loam: $K = 227 - 0.11V$
Clay: $K = 341 - 0.19V.$

Further research, which has been summarized by Young [7.4], showed that the form of Poncelet's relationship, as presented by Petry, would be improved if a factor were introduced to take account of the shape of the nose of the projectile, if penetration was taken proportional to the square root of the projectile pressure $(W_p/A)^{1/2}$, and if the relationship between penetration and V^2 was replaced at higher striking velocities with a relationship between penetration and V. On this basis, Young discussed the following equation for penetration into soil:

$$p = 0.53SN(W_p/A)^{1/2}\log_e(1 + 2V^2/10^5), \qquad V < 200 \text{ ft/sec},$$
and
$$p = 0.0031SN(W_p/A)^{1/2}(V - 100), \qquad V \geq 200 \text{ ft/sec}. \tag{7.5}$$

The units are reconciled in the constant terms 0.53 and 0.0031.

S is the soil constant and N a nose-performance coefficient, which takes account of the nose shape, ranging from a flat nose ($N = 0.56$), to a cone shape having a length of cone equal to three times the diameter ($N = 1.32$). Note that, contrary to the theories of Poncelet and Petry, penetration in soil is proportional to $(W_p/A)^{1/2}$ rather than W_p/A. Values of S and N were found experimentally, and linked to broad ranges of soil type thus:

Rock:	$S = 1.07$
Dense, dry silty sand:	$= 2.5$
Silty clay:	$= 5.2$
Loose, moist sand:	$= 7.0$
Moist clay:	$= 10.5$
Wet silty clay:	$= 40$
Soft wet clay:	$= 50$.

Thus, all other things being equal, penetration in soft wet clay is about 50 times as far as in rock, and about 20 times as far as in dense, dry, silty sand.

Typical values of N are:

Flat nose	$N = 0.56$
Tangent Ogive 2.2 CRH	$= 0.82$
Tangent Ogive 6 CRH	$= 1.00$
Cone ($L/D = 3$)	$= 1.32$.

Care must be taken not to give these figures a greater scientific accuracy than they merit, since the scatter of penetration test results is notoriously wide. Experimental data exists on penetration depths up to 220 feet and in Young's equations there is apparently no upper limit to the value of p for penetration into homogeneous soil. There is a lower limit, however, and it is suggested that the equations apply as long as the total depth of penetration is equal to 'three body diameters plus one nose length'. At lesser depths the mechanics of penetration are not fully activated. Further, if the nose length is more than one-third of the total penetrator length there is insufficient length of cylindrical section to ensure stability, because the centre of gravity of the penetrator is too far aft.

It is useful to note that the practical range of W_p/A is fairly limited, and ratios greater than 15 to 20 psi are difficult to achieve. For solid steel, to take an extreme example, a billet having a diameter 4 in and a length of 60 in, has a value of $W_p/A = 17$. There is also a practical range for velocity, V, in Young's equations. He suggests that at impact velocities less than 105 ft/sec, the penetration depth is too shallow for reliable analysis.

The range of validity of Young's equation was given by him as:

Weight (W)	2 to 3750 lb
Diameter (in)	1 to 30 in
W_p/A	0.1 to 38 psi
V	100 to 2370 fps.

Young also examined penetration into layered earth, by establishing equations to calculate the change in velocity of the penetrator as it completes penetration of the first layer, and then to use the velocity of exit as the new incident velocity for the second layer, and so on. The thickness of any soil layer must be greater than the length of the nose of the projectile for Young's simple analysis to apply. If this is not so, there are complications to the theory.

Goodier [7.5] suggested in 1965 that the pressure acting on a penetrating projectile was equivalent to the pressure of resistance exerted on an expanding spherical cavity. Later Norwood [7.6] proposed that a more realistic representation of the target material would be an expanding cylindrical cavity, and then Yew and Stirbis [7.7] assumed that the tunnel formed by the impacting projectile could be represented by a successive series of spherical cavity expansions. The radii of the final cavities are governed by the nose geometry.

Yew and Stirbis obtained the equations of motion of the projectile and from these predicted the penetration of a projectile into limestone. The results indicated that their theory gave close agreement with tests – closer than would have been estimated by the formula due to Young [7.4]. For a much harder material (weeded tuff) Young considerably overestimated penetration, whereas the theory of Yew and Stirbis again gave close agreement.

All the early penetration theories for the entry of projectiles into soils are special cases of the general relationship

$$-\frac{dV}{dt} = \alpha V^2 + \beta V + \gamma, \tag{7.6}$$

where γ is a positive constant. This has been pointed out by Allen, Mayfield and Morrison of the US Naval Ordnance Test Station, China Lake, California [7.8]. Poncelet's equation, for example, corresponds to Eq. (3.6), with $\beta = 0$, and α and γ positive constants. The earlier work of Robins [7.9] and Euler [7.10] took α and β equal to zero, and γ as a positive constant. Allen *et al.* related the coefficients of the penetration formulae to the drag coefficient used in the analysis of a projectile passing through a continuous fluid having a density ρ. If the projectile has a mass m, and a presentation area on a plane normal to the flight line of A, the analysis of fluid dynamics gives

$$-m\frac{dV}{dt} = \tfrac{1}{2}C_d A \rho V^2, \tag{7.7}$$

where V is the velocity of the projectile. Although C_d can vary with velocity, it is thought to be reasonable to assume that it remains virtually constant over a wide velocity range when the medium is granular rather than a fluid.

Reference [7.8] also summarizes the results of a test programme, in which steel projectiles, 0.511 in diameter, 5.11 in long, were fired from a Browning aircraft machine gun into dry quartz sand. The ends of the projectile were conical in shape, and a range of cone angles were investigated, from 180° (flat end) to 10° (very sharp). By means of a 0.404 in diameter hole drilled inwards from the base of the projectile, the centre of gravity could be moved forwards, and it was found that to ensure stability after entry into the sand the length of the internal hole should be such that the centre of gravity of the projectile coincided with the base of this hole.

A comparison of penetration versus striking velocity suggested that at lower velocities penetration would be sufficiently accurately described by the Poncelet equation, whereas at higher velocities conventional penetration theories did not predict the results adequately. For the higher range ($V > 96.5$ m/sec), the relationship was closer to the form

$$-\frac{dV}{dt} = \alpha V^2. \tag{7.8}$$

This gave a discontinuity in the relationship between velocity and penetration at the critical velocity (V_c) of about 100 m/sec.

It was also noted that C_d was relatively insensitive to cone angle (2ϕ) in the range $2\phi > 90°$, but at sharper angles the drag coefficient fell in value, becoming equal to 1.0 when $2\phi = 30°$. The beginning of the relatively sharp fall in C_d coincided with the achievement of the critical velocity (V_c), and this velocity appeared to be related to the velocity of sound in sand.

Values of C_d were given as follows:

(a) For a flat nosed projectile in a uniform stream of discrete particles, with perfectly elastic impact, $C_d = 4$.
(b) For a flat-nosed projectile ($2\phi = 180°$) in a uniform stream with perfect inelastic impact, $C_d = 2$.
(c) For the circumstances of the test, $1.97 > C_d > 1.80$ when 2ϕ varied between 180° and 100°. For $2\phi < 100$, $1.80 > C_d > 1.0$.

Artillery shells, and later bombs, were the subject of much testing, particularly during the run up to the Second World War, and during the early years of that conflict. The position in 1946 was summarized by Stipe [7.11] in the USA and by Christopherson [7.12] in the UK. The former pointed out that early wartime tests on projectiles and bombs were conducted over a limited range of velocities, and very often the dimensions and weight of the missiles, their striking velocity and the soil characteristics were not recorded. This was unacceptable to the Chief of Engineers, US Army, who asked for a full study of the terminal ballistics of soil to be made. The lack of information was also recognized in the

UK, and firing trials were organized by the Road Research Laboratory to investigate the scientific principles in a methodical way.

The US Report noted that similar projectiles would penetrate two to three times as far in a rich clay than in coarse sand, and that soil stratification could affect the depth of penetration. Penetration into cohesive soils formed conical craters, wide at the entrance and gradually tapering for the length of the penetration path. Displaced soil was compacted, and sometimes pulverized. It was noted that projectiles striking the soil surface at high angles of obliquity often followed a curved path, and that aerial bombs sometimes took a J-shaped trajectory underground, curving forward in the direction of the flight line of the aircraft. There was a tendency for projectiles to be unstable in end-on motion, so that they 'tumbled' into a side-on attitude, which often curved the trajectory. Projectiles came to rest with their noses pointing back to the point of entry, and blunt nosed missiles were found to 'topple' less quickly than sharp-nosed varieties.

The American work was summarized in the form of data sheets, which eventually found their way into the US Army technical manual 'Fundamentals of Protective Design', TM5-855-1 (7.13). From this manual we have taken Figure 7.1, which gives the correction term to be applied to calculated vertical penetration when oblique penetration occurs. This term is a function of the striking velocity, V, and it can be seen from the figure that if the cosine of the angle of obliquity were used as the correction term, the vertical component would be overestimated.

On the assumption that the J-shaped penetration path was a valid approximation, its length was given as part of weapon data in ref. [7.13]. This indicated that, in general, the straight part is about two-thirds of the total length and the radius of the curved part is between one-fifth and one-third of the total length.

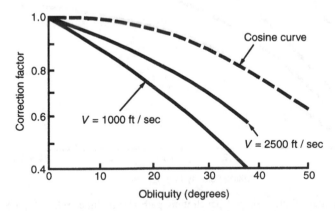

Figure 7.1 Correction factor v angle of obliquity for two values of striking speed (from Stipe, ref. 7.11; US Army TM5-855-1, 1965).

The penetration length, L, which is the total length of the J-path, not the final vertical depth, p, was found from the tests to be linked to striking velocity through the parameter $L/W_p^{1/3}$, where W_p is the total projectile weight, not the weight of the explosive charge (W). The curves shown in Figure 7.2 are taken from ref. [7.13], and refer to four types of soil and projectiles having an 'average' nose shape. Detailed information on other nose shapes are given in the reference.

The vertical penetration of a range of general purpose bombs, with total weights in the range 100 to 2000 lb, in three different soils, is given in Figures 7.3, 7.4 and 7.5. The curves were said by Stipe to agree with the scatter of available data to within ±20%, and a rough interpolation between curves could be made for other soil types. Because the tests were mainly carried out with free-dropped aerial bombs, the striking velocity has been replaced by the altitude from which the bombs were dropped in level flight. In each experiment, delay fusing prevented explosion until full penetration had been achieved, and the

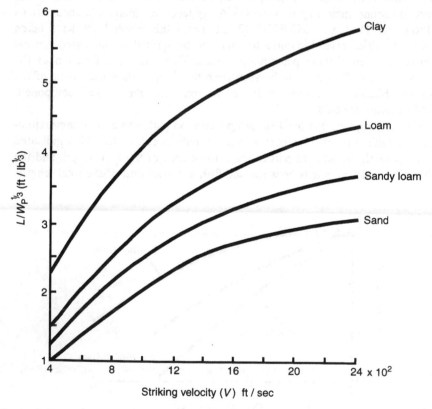

Figure 7.2 Length of the J-shaped penetration path (from Stipe, ref. 7.11; US Army TM5-855-1, 1965 and 1986).

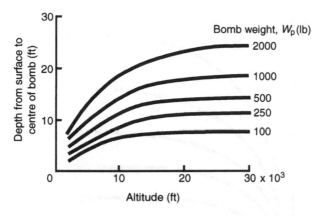

Figure 7.3 Penetration of US GP bombs in sand (from Stipe, ref. 7.11).

results confirm the earlier statement that penetration into clay is about double that into well-compacted sand. More recently, doubt has been cast on the assumption of a J-path, and on the relevance of analysis based on this concept.

It was realized that the stability of projectiles in soil needed further investigation, and since stable underwater rockets had been developed during the Second World War, it was suggested that these might also be stable in earth. Whether this line was ever followed is not clear now.

In the UK, Christopherson's report [7.12] was as always a very precise and valuable analysis of the fundamentals. He classified the soils into three types,

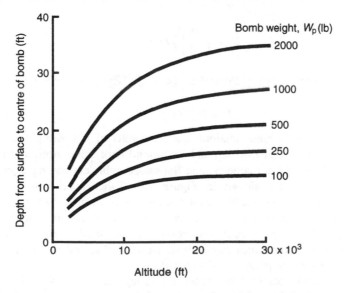

Figure 7.4 Penetration of US GP bombs in loam (from Stipe, ref. 7.11).

Penetration and fragmentation

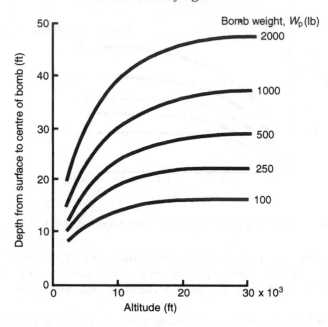

Figure 7.5 Penetration of US GP bombs in clay (from Stipe, ref. 7.11).

plastic materials like clay, granular materials like sand or gravel, and the softer rocks such as chalk and sandstone. He assumed that for soil penetration the variables were separable, so that the relationship between total track length, L, the diameter of the projectile, d, and striking velocity, V, could be written in the form

$$\frac{L}{d} = \left(\frac{W_p}{d^3}\right)^n f(V). \tag{7.9}$$

If the velocities given to the material are largely independent of diameter, and if $n = \frac{1}{3}$, then

$$L = W_p^{1/3} f(V). \tag{7.10}$$

Using this as a governing relationship, the results of UK firing trials into undisturbed natural clay at Dorking and Richmond Park using armoured piercing shot, steel balls of $\frac{1}{2}$ in diameter, and flat-ended cylinders having diameters of 0.50 and 0.78 in, are shown in Figure 7.6. Either maximum or mean penetration times are given, but the results in general confirmed the large scatter associated with soil experiments. Christopherson suggested that missile instability caused the wide scatter, since it was known that almost all the missiles turned over at least once – sometimes more often. He also observed that penetration increased as the average grain size decreased.

The British Bomb Disposal Units recorded during the Second World War the depths from which bombs were recovered, and these were plotted in the form

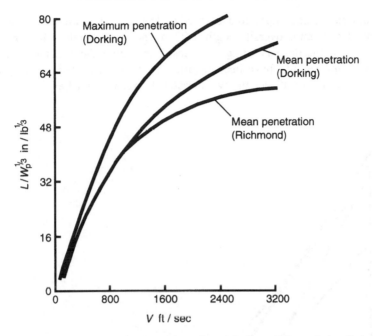

Figure 7.6 Penetration track length in clay of steel balls and flat-ended cylinders (UK Road Research Lab., 1944; Christopherson, ref. 7.12).

shown in Figure 7.7. The relationship was given in terms of vertical penetration, p, divided by the cube root of bomb weight ($W_p^{1/3}$), and the number of bombs per thousand recovered that were found at greater depths than that indicated. The mean penetration (exceeded by half the bombs (500 out of 1000)) was given by $p/W_p^{1/3} = 2.76$, and the average striking velocity was 850 ft/sec. There was a 99% certainty that, in spite of scatter, penetration of German bombs never exceeded $p/W_p^{1/3} = 5$.

In dry sand no permanent crater was visible at the entrance hole, and the projectiles usually turned side-on before coming to rest. The mean track length at 850 ft/sec initial velocity was 2.33 $W_p^{1/3}$, and the mean vertical penetration was 1.95 $W_p^{1/3}$, giving a p/L ratio of 0.8.

In chalks and sandstones it was found that track length and penetration are about equal. The best fit to experimental results was to take n in Eq. (7.9) as $\frac{2}{3}$, and curves on this basis are given in Figures 7.8 and 7.9. It was generally felt that by 1946 a much clearer idea of bombs and shell penetration existed, but much research remained to be done. It is doubtful whether much has been accomplished of equal scientific importance since those days, although much analytical work has been attempted.

If we may include pack-ice as a geological target, then it is relevant to refer to test work in the late 1960s and early 1970s by the Sandia Laboratories. Four penetrators with conical noses were air dropped on to pack-ice in the north-west territory of Canada, and impacted at a speed of 159 m/s. Their dimensions were: outer diameter 70 mm, overall length 1.07 m and nose length 140 mm; the weight of each missile was 23 kg. The tests, including the measurement of deceleration/time records, have been reported by Young and Keck [7.14], and compared with analytical models by Forrestal, Longcope and Lee [7.15]. The

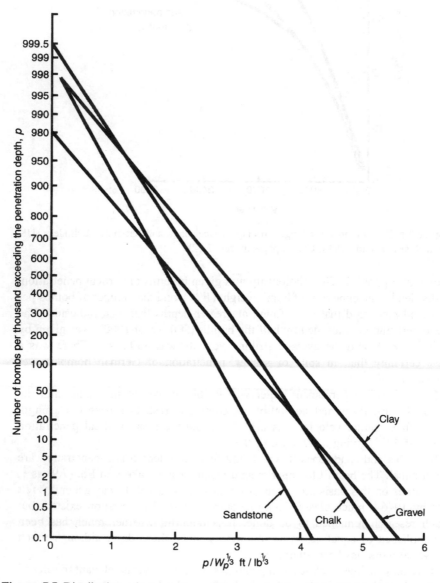

Figure 7.7 Distribution of penetration of German bombs (from Christopherson, ref. 7.12).

latter had earlier used elastic-plastic analysis to develop theories for the prediction of forces on conical nosed penetrators, in which the cylindrical cavity approximation, put forward originally by Bishop, Hill and Mott [7.16], was used. (See also ref. [7.6].)

A similar analysis was applied to tests in which penetrators having a total length of 1.5 m, a diameter of 0.156 m, an ogival nose with 6.0 CRH and a mass of 162 kg, were fired at a dry porous rock known as Antelope Tuff. Data from the earlier work of Young [7.17] suggested that the ogival nose and a conical nose with half apical angle $\phi = \tan^{-1} 0.30$ were about equivalent. The penetrators were fired from a Davis gun and impacted the rock at a speed of 520 m/s, as described in a report by Longcope and Forrestal. It was noted that sliding friction helped to increase penetration resistance.

Further work was carried out by two of the authors of ref. [7.15] at high impact velocities by firing simulated soft sandstone targets at penetrators, using a gas gun. Striking velocities between 0.2 and 1.2 km/s were obtained on impact with 20.6 mm diameter penetrators, and it was found that over a range of

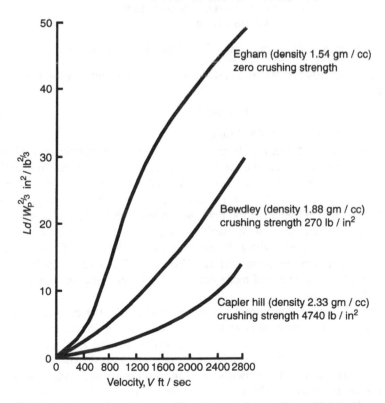

Figure 7.8 Penetration of small projectiles into sandstone (from Christopherson, ref. 7.12).

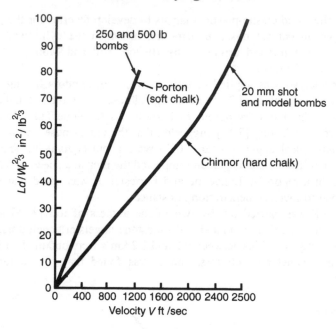

Figure 7.9 Penetration of projectiles into chalk (from Christopherson, ref. 7.12).

striking velocities between 500 and 1200 km/s, the axial resultant nose force (F) was given by

$$F = 109V^{1.29}, \qquad (7.11)$$

where F and V have units of KN, km/s.

In the late 1970s, the study of fragment ballistics led to experiments on the penetration mechanics of steel fragments into sand, because sand was frequently used as a protective medium around buildings and hardened structures against flying steel projectiles. The tests were reported after a gap of some years by Stilp, Schneider and Hülsewig of the Ernst Mach Institute [7.18], and concerned penetration depths associated with idealized fragments in the form of spheres and cylinders. Steel spheres with a diameter of 1 cm and a mass of 4.07 g were shot into dry loose sand by Schneider and Stilp in 1977, at velocities between 118 m/s and 3553 m/s. The sand had a density of 1.8 g/cm³ and maximum grain sizes of 1.05 and 2 mm.

It was noted that for low impact velocities (<400 m/s) the projectiles were decelerated by sand displacement and compaction, and the kinetic energy was dissipated as internal friction. At velocities >400 m/s the sand grains also became fragmented and the projectile material was eroded. The amount of erosion was linked to the hardness of the projectile steel. At velocities >2000 m/s the projectiles broke into two or three parts. It was also concluded that during the penetration process the projectile remained at a constant orientation, with a cavity-like transient penetration channel formed behind it.

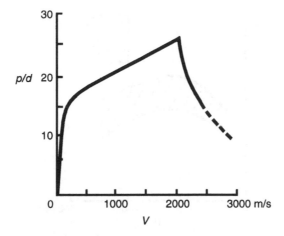

Figure 7.10 Penetration curve for spheres into dry sand (from Stilp *et al.*, ref. 7.18).

This behaviour is shown in Figure 7.10, which relates penetration/diameter (p/d) with the velocity of impact (V) in m/s. The curves, which are the mean lines of experimental results, indicate a sharp rise in penetration at low velocities followed by a slower rate of increase. When the projectiles break up there is an increase in scatter and a sudden drop in penetration depth. After each test the projectile was weighed to assess the mass loss due to erosion. Starting at a velocity of 500 m/s, the mass loss increased linearly with impact velocity, as shown in Figure 7.11. The mass loss at 2000 m/s was about 4%.

Cylinders having length/diameter ratios of 1, 5 and 10, and masses between 10 g and 50 g, were shot into sand at velocities between 93 m/s and 1400 m/s by Hülsewig and Stilp. There was a noticeably larger scatter in results than for

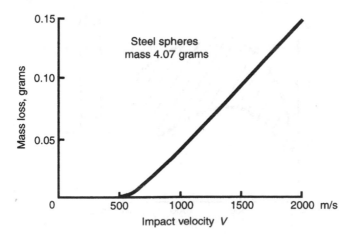

Figure 7.11 Absolute mass loss of steel spheres on impact with sand (from Stilp *et al.*, ref. 7.18).

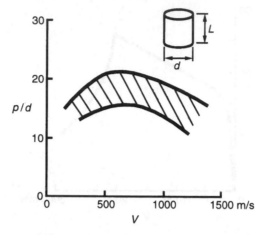

Figure 7.12 Penetration experiments on cylinders, $L/d = 1$ (steel cylinders into sand) (from Stilp *et al.*, ref. 7.18).

spheres, partly because the cylinders became flattened and 'mushroomed' at the higher impact velocities. The greater the length/diameter ratio, the greater the scatter, and at ratios of 10 the projectiles bent on impact. This caused deviations from the original trajectory, made worse when the cylinders also broke into pieces. Values of p/d for cylinders with $L/d = 1$, 5 and 10 are shown in Figures 7.12, 7.13 and 7.14.

The prediction of projectile penetration into rock and cemented soils by means of a numerical procedure was examined by Schoof, Maestas and Young, who reported their work in 1989 [7.19]. The Sandia National Laboratories at

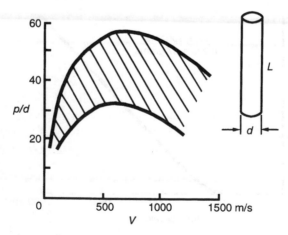

Figure 7.13 Penetration experiments on cylinders, $L/d = 5$ (steel cylinders into sand) (from Stilp *et al.*, ref. 7.18).

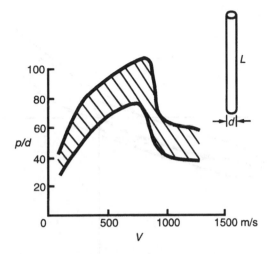

Figure 7.14 Penetration experiments on cylinders, $L/d = 10$ (steel cylinders into sand) (from Stilp *et al.*, ref. 7.18).

Albuquerque had developed a computer code (SAMPLL), which was a simplified analytical model of penetration and lateral loading, and this was updated to include hard soil and rock empirical equations. The code was based on the penetration equations developed by Young, which were discussed earlier, but were not at the time suitable for predicting soft soil penetration. Figures 7.15 and 7.16, taken from the 1989 paper, show the notation, which includes angle of attack, trajectory angle and velocity vector, and show how the paths of a penetrator can be simulated.

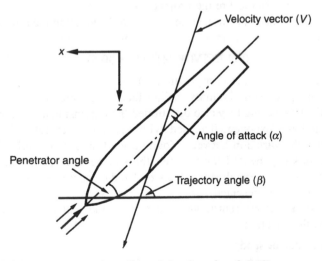

Figure 7.15 SAMPLL conventions (from Schoof *et al.*, ref. 7.19).

Penetration and fragmentation

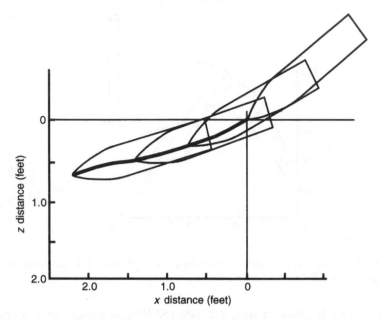

Figure 7.16 Penetration simulation by SAMPLL (from Schoof *et al.*, ref. 7.19).

A further investigation of the accuracy of Young's equation was reported in 1989 by Taylor and Fragaszy [7.20], who used a centrifuge to carry out soil penetration tests at accelerations greater than 1 g. Spherical projectiles were shot into clean quartz sand which was controlled tightly on density and uniformity. The projectiles were made of brass, with diameters of 5.69, 7.82, 9.09 and 10.92 mm, and entered the sand at velocities of about 300 m/s. A Thompson Contender pistol was used to fire the projectiles.

The authors questioned several features of Young's equation (see Eq. (7.5)), which, for velocities greater than or equal to 61 m/s, can be written as

$$p = 0.117 KSN(W_p/A)^{0.5}(V - 30.5) \qquad (7.12)$$

(when p is in cm, W_p in kg, A in m^2, V in m/s).

They were unhappy about the mass scaling factor, K, which they considered had no theoretical basis, and they also questioned the dimensional correctness of the equation and the exponent 0.5 used with the mass/area ratio. We have discussed earlier the variation between Young's work and other early theories with regard to this exponent. They also felt the soil penetration index, S, was rather qualitative – a problem already understood and discussed at the time when Young's work was first presented.

As a result of about 80 centrifuge model tests, the authors produced the following 'best fit' equations:

Dense Ottawa Flintshot sand,

$$p = 0.00282(W_p/A)^{0.915}, \qquad (7.13)$$

Loose Ottawa Flintshot sand,

$$p = 0.00634(W_p/A)^{0.88} \tag{7.14}$$

The values of S, buried in these formulae, were given as 0.137 for dense sand, and 0.30 for loose sand. Note that the exponents are considerably greater than the 0.5 originally given by Young. The authors went on to investigate whether all tests could be accommodated by the use of one formula, if the W_p/A term were replaced by a dimensionless parameter. They proposed π_{Dr}, where

$$\pi_{Dr} = (M_p g)/(A_p l_t V^2 Dr^{0.5}). \tag{7.15}$$

M_p and A_p are the mass and area of the projectile, l_t is the total soil density, and Dr is the relative density of the soil, which is a measure of the density in relation to its minimum and maximum values.

The equation for penetration depth in metres for the Ottawa sands then becomes

$$p = 21.427(\pi_{Dr})^{0.894} \tag{7.16}$$

(p in metres).

In recent years there has been considerable interest in the use of rock rubble to help limit the penetration of aerial bombs and missiles, when these are aimed at underground facilities. At the time of writing much of the work has a security classification, certainly in terms of performance figures, but it has been generally accepted that layers of rock rubble give a very high protection against penetration by general purpose bombs, but are less effective against armour-piercing bombs. A 20 foot thick rock rubble field is reckoned to be an effective countermeasure against advanced designs of penetrating weapons, but this is a very expensive option in terms of raw material and excavation. There is some doubt whether conventional air bases are fully protected in terms of penetration, and it is thought they could be vulnerable to advanced weapons.

Recently an analytical method of solving the problem of penetration through rock rubble has been put forward by Gebara, Pau and Anderson [7.21]. The authors make the point that although rock rubble overlays are considered a good alternative to brick and expensive burster slabs for protecting underground installations from high kinetic energy penetration projectiles, the modelling of rubble systems has proved difficult. The Waterways Experiment Station of the US Army have devoted research funds to the problem, as illustrated by the work of Nelson, Ito, Burks et al. [7.22] and Gelman, Richard and Ito [7.23]. The latter took the rubble as a uniform mesh of octagons, which is a rather simplified fracturing geometry. Ref. [7.21] presents a solution using the Finite Block Method, put forward earlier by Chen and Pau [7.24], in which the penetrator may burrow its way into the mesh of blocks without fragmentation, or where the penetrator, on impacting one block, may cause it to fragment. In the latter case the Voronoi construction for fragmentation was chosen, in which the block area is partitioned by perpendicular bisectors between random points. This produces

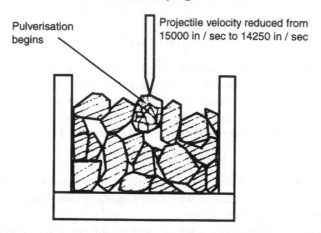

Pulverisation
begins

Projectile velocity reduced from
15000 in / sec to 14250 in / sec

Figure 7.17 The process of fragmentation and penetration into rock rubble by the finite
block method (from Gebara *et al.*, ref. 7.21).

fragments that are reasonably equal in length and breadth, rather than heavily
elongated, and is thought to be well representative of experiments. Figure 7.17,
taken from ref. [7.21], shows a penetrator hitting a single block, causing it to
fragment. The size of rocks in a rock rubble fabrication is about three times the
diameter of the projectile, so the fragmentation of individual rocks is highly
likely. This results in less deflection from the direction of the projectile path than
would be the case in a non-fragmenting block analysis.

Earlier mention of the SAMPLL computer code is a reminder that considerable efforts have been made in recent years to establish computer software for
the prediction of penetration into soils and other materials. In 1987 a paper by
Schwer, Rosinsky and Day [7.25] presented a computational technique for earth
penetration, which included direct coupling between a deformable target and a
deformable penetrator. In their review of recent computational work they called
attention to the code PENCO, which was established at the US Army Waterways Experiment Station in the early 1980s, and reported on at the time by
Creighton [7.26]. This code treats multi-layered targets of hard or soft rock
materials, where the target resistance is specified by the unconfined compressive
strength, or by the S factor in Young's equations (see Eq .7.5). It does not treat
the interaction between penetrator and target for deformable penetrators, and
appears to be limited in its treatment of very thin targets having thicknesses of
only one or two penetrator diameters. Reference was also made to the possible
use for soil penetration, using rigid penetrators, of the codes HULL [7.27] and
TRIFLE [7.28], which were Eulerian-based finite difference codes developed
mainly for fluid dynamics problems.

The Lagrangian penetration grid consists of a narrow tunnel that coincides
with the line of the penetrator trajectory, and a typical target grid, taken from
ref. [7.25] is shown in Figure 7.18. The use of the tunnel is to reduce

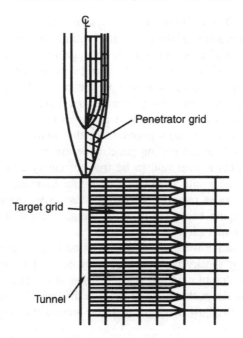

Figure 7.18 Typical Lagrangian penetration grid with narrow tunnel passing through the target (from Schwer *et al.*, ref. 7.25).

computation time by eliminating the analysis of highly deformed material directly in front of the penetrator nose, and the tunnel cross-sectional area is often made equal to the small flat area in the end of the nose. The concept of material moving radially away from the penetrator as the penetration takes place is the theoretical basis for the codes PENCO and SAMPLL, and the assumption is made that the kinetic energy of the penetrator is transferred to the strain-energy of radial expansion of the target material. The presence of a small tunnel is not reckoned to have a noticeable effect on the energy transference.

The analysis of projectile penetration through a narrow cylindrical drill in soil has been presented by Yankelevsky [7.29], who represented the soil medium as a set of thin discs. Expressions for the interaction pressure between the discs were derived, and all drill sizes between zero diameter and projectile diameter were examined.

The introduction of a three-dimensional analysis of penetration gave rise to the code EVA-3D, and this has been mainly used in conjunction with concrete slabs. The code SAMPLL has also been extended into 3D situations (SAMPLL-3D), as described by Young and Young [7.30] in their analysis of a simple model of penetration and lateral loading. In all these systems, the input is the penetrator geometry (shape and mass) and the target geometry in 3D form. From these the finite element preprocessor generates the mesh, a material model is established and the initial impact conditions are set up. The analysis, by

EVA-3D, SAMPLL-3D or a Finite Element system, leads to the generation of time histories and trajectories of the penetrator within the target material.

The modelling of projectile penetration into geological targets has been investigated by Amini and Anderson [7.31], using energy tracking to satisfy the kinematics in the target material adjacent to the projectile, and momentum/impulse to track the momentum of the projectile as a function of time. Particular attention was given to projectiles that strike at angles different from 90° to the target surface, because the authors pointed out that many of the empirically based analytical methods of calculating penetration are only reliable for normal directions of strike. This is also said to be true for cavity expansion theories. Numerical models using Finite Difference or Finite Element general purpose mainframe computer codes require expensive constitutive information on the target material and require excessive time for the generation of input files and the interpretation of output files.

The stages of penetration were taken in four parts: local material degradation due to the propagating shockwave in the vicinity of the first impact; partially restrained penetration due to chipping and cratering at the point of impact; the hitting of the target by the pusher plate or sabot at the end of the projectile when it is fully embedded; the flow of target material around the projectile into the wake cavity behind it.

The radius of failed material due to initial impact was calculated, and the properties of material within this radius were used to calculate the resisting forces on the projectile. These forces were found by integrating static and dynamic stresses over the contact area of the nose and body, and were then used to calculate depth, velocity and deceleration of the projectile as a function of time for each time increment. The results of the analysis were compared with two experiments in which the deceleration of projectiles into Tuff limestone rock were measured. The comparison between model prediction and experimental results for Thirsty Canyon Tuft is shown in Figure 7.19. Readers wishing to examine more closely the characteristics of wedge penetration into rock should consult a paper by Pariseau and Fairhurst [7.32].

The capability of civil and military structures to withstand rifle fire is an important area of research, particularly when the resistance can be increased by the judicious use of lightweight cladding. A study was completed in 1983 on behalf of the military engineering research establishment of the UK Ministry of Defence by Anderson, Watson, Johnson and McNeil [7.33] in which the penetration of high-velocity projectiles into rock/polymer composites was examined.

The old-fashioned way of rapidly protecting buildings was by sand bags, but they tended to be impermanent and bulky. A form of rapidly assembled cladding was therefore a preferred solution, providing that a cheap and effective design could be found. Apparently it was found during the Second World War that dry gravel in a wire mesh box gave useful protection, and a step forward from this was to cast the gravel in a cold cure rubber binder. Rubber can be expensive, so

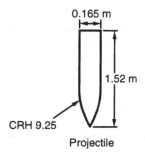

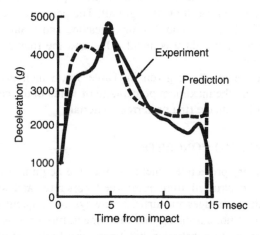

Figure 7.19 Penetration data for a projectile entering Thirsty Canyon tuff (from Amini and Anderson, ref. 7.31).

the possibility of using commercially available cheap polymers was investigated.

In analysing the mechanics the authors made use of the empirical equations given in the 1972 report by Young [7.4]. Most analytical methods are aimed for simplicity at homogeneous materials, where the projectile cross sections are orders of magnitude greater than the size of the aggregate. This simplification does not apply to aggregates bonded by a low-elasticity matrix of polymer with sand filler, particularly when the projectile sections are similar in dimension to the size of the aggregate, and when the projectile strikes the matrix at an oblique angle.

Anderson *et al.* reported a series of tests, in which hardened steel core projectiles, 7.62 mm diameter, 33.3 mm long, were fired at various mixes of polymer and rock aggregate. Since impact trajectory through the composite was not straight, it was necessary to distinguish between the total length of the

Table 7.1

	Gravel	Basalt	Limestone
Block thickness (mm)	100	110	120
Mean penetration (mm)	51	81	75
Max % voids after compaction	47	50	48

Striking velocity: 807 m/s with standard deviation of 9.8 m/s

penetration path length and penetration normal to the face of the composite specimen. The tests were carried out with either Polyester Polyurethane or a blend of Polyether Polyurethane as the base material, and with aggregates of crushed limestone, crushed basalt or river gravel. The percentage of polymer by weight was about 9% for gravel and 7% for limestone and basalt. The percentage of rock aggregate by weight was about 60%, and the rock size was in the range 26.5 to 37.5 mm.

An indication of the results is given in Table 7.1, which gives the mean normal penetration, and the maximum percentage of voids in the cast specimens for the Polyether Polyurethane (the preferred material).

7.3 PENETRATION INTO CONCRETE

Since most new military protective structures since the beginning of the present century have been constructed from reinforced concrete, and since military targets such as buildings, bridges and airfield runways are frequently constructed from this material, it is not surprising that the expenditure on research to study the response of concrete to penetration has been very high. Let us take as our starting point the scientific research of the Second World War, starting as always with the work of Christopherson [7.12].

Much of the wartime research was concerned with the penetration of aerial bombs into reinforced concrete slab-type structures, so it became necessary to classify bombs in terms of their penetrative power. Four types of bombs were recognized in Britain, and the first of these was armour-piercing (AP) bombs, similar to naval shells, with low charge/weight ratio (10–15%) and relatively high-length diameter ratio. Weights of 1000 lb and above were registered. Although 'armour piercing' suggests metallic targets like battleships or tankers, they were also used against protective concrete. The second type was 'semi-armour-piercing' (SAP) bombs, with a higher charge/weight ratio (20–30%), and lower weights, in the range 250 to 1000 lb. The third type was demolition bombs, also known as medium capacity or general purpose bombs. Charge/weight ratios were in the range 45–50%, and weights covered a long range from 250 lb upwards. Their damage effect was about double that of SAP bombs, which was not surprising since they carried about double the charge weight. The last category was the high capacity bombs, with a charge/weight ratio between

70 and 80%, and a weight range starting at 2000 lb and going upwards. They destroyed buildings by blast and ground shock rather than by penetration and explosion.

The fusing was important to the type of loading that was imparted to targets. Medium capacity bombs fused instantaneously, so that no penetration occurred before detonation, and these were used when the fragmentation of the casing was required to be the major threat – particularly against soft skinned vehicles or against the light metallic structures of aircraft. Fragments were not thought to be very important when general damage to surface structures was required.

In addition to the above categories, very large bombs were designed for use against particular targets. Examples that received much publicity were the 'Tallboy' bomb (12 000 lb) designed to attack reinforced concrete German U Boat pens on the Atlantic coast of France (and also used against V-weapon sites), and the Grand Slam bomb (22 000 lb). Barnes Wallis (much later to be knighted) was clear that it was necessary when attacking important buried targets to 'inject the largest possible explosive charge to the greatest possible depth in the medium'. In spite of the accent on penetration, the first Grand Slam bomb was in fact dropped in March 1945 in operations against the Bielefeld Viaducts in Germany. The first Tallboy bomb was dropped in June 1944, and it is interesting to note that near misses up to 40 feet were more damaging than direct hits, particularly when attacking concrete, brick or masonry bridges and viaducts. The extensive damage was caused by the effect of earth shock on the foundations.

Forty-one Grand Slam bombs were dropped in the last two months of the war, with the following penetration figures: Height of drop 16 000 to 18 000 feet, penetration into sand or chalk 60 to 75 feet, penetration into clay 90 to 100 feet. The Tallboy bombs, dropped from the same height, penetrated to depths about two-thirds as deep.

For a given striking velocity of 1000 ft/sec, the average depth of penetration of these bombs was linearly related to the cross-sectional density of the bomb (W/A) in lb/in², as indicated in Figure 7.20. The Tallboy bombs used to attack U Boat pens penetrated the reinforced concrete roof by more than 6 or 8 feet before exploding, and subsequent blast from the exploding charge caused a further 10 feet of penetration. The general view of the time was that a Tallboy bomb was completely effective against capital ships, viaducts and large bridges, tunnels with up to 50 feet of overburden, and reinforced concrete roofs 10 feet thick.

From the extensive range of wartime data on the penetration of all types of bomb into reinforced concrete, the British Road Research Laboratory produced an empirical formula, as follows:

$$p = 6.3\left[1 - \frac{\sigma_c - 2500}{20\,000}\right]\frac{W_p}{d^2}\left(\frac{d}{c}\right)^{0.2}\left(\frac{V}{1000}\right)^{1.5}, \quad \text{inches,} \quad (7.17)$$

where

$$p = \text{penetration (in)}$$
$$W_p = \text{projectile mass (lb)}$$
$$d = \text{projectile diameter (in)}$$
$$\sigma_c = \text{crushing strength of concrete (lb/in}^2)$$
$$V = \text{striking velocity (ft/sec)}$$
$$c = \text{maximum aggregate size (in)}.$$

Later, after extended investigations involving larger ranges of diameter and aggregate size, and after noting that the power to which v was raised was a function of crushing strength, a final version of the UK equation, due to Whiffen (7.34), became

$$p = \left(\frac{870}{\sigma_c^{0.5}}\right)\left(\frac{W_p}{d^2}\right)\left(\frac{d}{c}\right)^{0.1}\left(\frac{V}{1750}\right)^n, \qquad (7.18)$$

where $n = 10.7/\sigma_c^{0.25}$.

The experimental ranges on which the equation was based were: σ_c: 800 to 10 000 lb/in^2; W: 0.3 to 22 000 lb; d: 0.5 to 38 in, d/c: 0.5 to 50; V: 0 to 3700 ft/sec; ogival projectiles with nose shapes between 0.8 and 3.5 calibre radius. The formula filled experimental data within a scatter band, slightly less than ±15%. The experiments indicated that when $V = 1750$ ft/sec the depth of

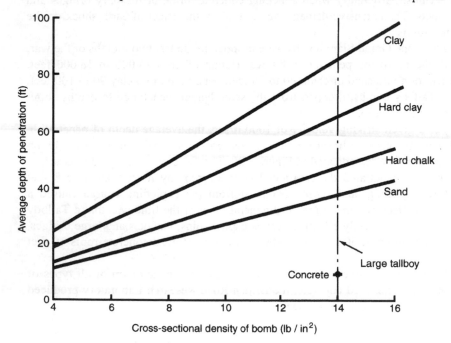

Figure 7.20 Average depth of penetration in feet for Tallboy bombs (from Christopherson, ref. 7.12).

penetration was inversely proportional to $\sigma_c^{0.5}$ and independent of n, and this explains the use of the $V/1750$ term.

The relationship between penetration and striking velocity for a projectile having $W_p/d^2 = 1$, $d/c = 1$, and $\sigma_c = 5000$ lb/in² is shown in Figure 7.21. The time of penetration into thick concrete slabs can also be found from the analysis, and is given by

$$T = \left(\frac{72.5}{\sigma_c^{0.5}}\right)\left(\frac{W_p}{d^2}\right)\left(\frac{d}{c}\right)^{0.1}\left(\frac{V^{n-1}}{1750^n}\right)\left(\frac{10.7}{10.7 - \sigma_c^{0.25}}\right), \qquad \text{seconds.} \qquad (7.19)$$

The American wartime work in this field was reported by Beth [7.35]. Small- and model-scale experiments were carried out at Princeton University, and large-scale tests were made at the Aberdeen Proving Ground. A number of National Defence Research Committee reports during the period 1942–45 contributed towards the final paper by Beth, in which he presented a penetration theory leading to the equation

$$p = \frac{266}{\sigma_c^{0.5}} \cdot \frac{W_p}{d^{1.785}} \left(\frac{V}{1000}\right)^{1.5} + \Delta l, \qquad \text{inches.} \qquad (7.20)$$

The term Δl was a nose correction, which took account of the assumption that the total volume removed by the penetration was more important than the length of the penetration path. For a flat-ended projectile $\Delta l = 0$, but for a spherical-ended projectile $\Delta l = d/6$.

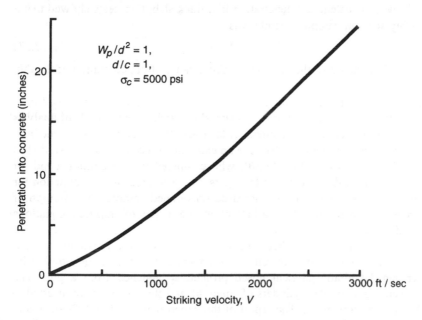

Figure 7.21 Penetration of projectiles into concrete (from Whiffen, ref. 7.34).

It was pointed out by Christopherson that whereas the British equation [7.18] was dimensionally consistent, the American formula incorporated a scale effect. When the work reported by Beth was later incorporated in the US Army Technical Manual on the Fundamentals of Protective Design [7.13], the formula had been further developed in the form

$$p = \frac{222}{\sigma_c^{0.5}} \left(\frac{4W_p}{\pi d^2}\right) d^{0.215} \left(\frac{V}{1000}\right)^{1.5}, \quad \text{inches.} \quad (7.21)$$

The units were similar in dimensions to those in the British formula, and as in that formula, the accuracy was said to be within ±15%.

During the Second World War it was realized that penetration into concrete was affected by the amount of cratering that took place at the point where the projectile first made contact with the slab, and by the amount of rear face spalling that might occur when the slab was relatively thin. The penetration formulae so far discussed assumed that the slab thickness was much greater than the penetration length of the projectile.

The front face spalling or cratering was estimated from the results of tests as follows:

$$\text{crater depth (in)} = 2.75 \times \text{projectile diameter } (d)$$

$$\text{crater diameter (in)} = 11.6d \, (c/d)^{0.125}, \quad (7.22)$$

and the rear face spalling or 'scabbing' was estimated for high explosive bombs and shells in terms of the minimum thickness of slab (t_p) to resist full penetration. If p was the calculated penetration in a thick slab, then tests showed that a reasonably accurate simple formula was

$$t_p = \tfrac{4}{3} \cdot p. \quad (7.23)$$

If t_p was less than this, the rear face spalling caused full penetration of the thin slab to occur.

The analysis of projectile penetration into mass-concrete or reinforced concrete structures continues to attract sponsored research. The methods of analysis vary from entirely empirical to attempts to model the behaviour of the medium using finite blocks or elements. To take an example, a recent empirical study has been made by Forrestal *et al.* [7.36], who examined the penetration depth of ogive-nosed projectiles into concrete targets, and who presented an equation for penetration depth in terms of an increased unconfined compressive strength of the concrete. In their work they call attention to a review of empirical equations presented earlier by Brown [7.37].

A second recent study has been made by members of the Picatinny Arsenal (US Army) on the resistance of reinforced concrete structures to penetration by Copper and Titanium spherical nosed projectiles launched from a gun. The targets were right circular cylinders of concrete, and steel plates were inserted to simulate the reinforcement. The report of the experiments by Gold, Pearson and Turci [7.38] stated that the penetration performance was analysed along the lines of Tate's theory for the deceleration of long rods after impact (Tate [7.39]. The essential feature of this analysis was that the erosion of the penetrator material takes place in combination with the motion of the penetrator as a rigid body. The range of projectile dimensions in the tests was between 1.27 and 2.00 cm diameter, 7.66 and 19 cm length. The impact velocities varied in the range 0.16 to 0.19 cm/μs. For the whole range of tests on both Copper and Titanium projectiles in plain or reinforced concrete, the depth of penetration was in the range 25 to 41 cm. Penetration into plain concrete was about 10% greater than into reinforced concrete.

The stress pulse initiated in cementitious materials as penetration occurs from an armour-piercing projectile has been investigated recently by Peters, Anderson and Watson [7.40] using a large diameter Hopkinson bar. There has also been an investigation of the penetration of a cylindrical steel projectile into fibre reinforced concrete plates by Dancygier and Yankelevsky [7.41]. The projectiles and conical hardened heads, were 75 mm long and weighed 118 g, and were accelerated to velocities of about 145 m/sec. The targets were either 40 or 60 mm thick, and there were several types of reinforcement. In general the penetration into the thicker targets was between 30 and 44 mm, and the 40 mm thick targets were completely penetrated.

7.4 PROJECTILE DAMAGE TO METAL STRUCTURES

It is well known that if a ductile metal plate of relatively soft material is penetrated by a non-deforming projectile, the material is pushed to one side following plastic deformation. A clean hole is made and 'petals' are formed around its circumference. As plate hardness is increased and the nose of the projectile is made more blunt, a plug may be formed at the head of the projectile. This plug is eventually ejected from the back face of the plate, and the plug varies in diameter from about one-third to the full diameter of the projectile. If the plating is poor in quality, it is possible that thin circular discs or flakes may be thrown off the back face. These flakes would normally be larger in diameter than the projectile.

Typical non-deforming projectiles (or at least with extremely low deformation due to impact) were the uncapped armour-piercing projectiles developed in the earlier part of the twentieth century. It was found that the projectile velocity at impact (V) that would be required if the projectile were to pass completely through a homogeneous steel plate was linked to the missile diameter (D), the plate thickness (t) and the total weight of the projectile (W_p). During the Second World War research showed that a 1000 lb bomb, 12 inches diameter, striking with a velocity of 800 ft/sec and having a value of W_p/D^3 of about 0.6 lb/in², would be able to penetrate completely between 4.5 and 6 inches of armour. Experiments had a fair scatter, but in broad terms there was an approximately linear relationship between t/D and $2V^2W_p/D^3$, at the upper limit of scatter, and a power relationship at the lower limit. For the perforation of metal plates with a Brinell Hardness Number between 250 and 300, and a projectile striking with zero obliquity with t/D in the range 0 to 2.0, the lower scatter band relationship was found to be $W_p V^2/D^3 = (t/D)^{3/2}$.

Early publications on metal penetration, taking account of the plastic deformation of the projectile were due to G. I. Taylor [7.42, 7.43]. An approximate theory of armour penetration, taking account of energy dissipation as the result of plastic deformation, and also considering the heating of the interface between the projectile and the plate, was established by Thomson [7.44] in 1955. During the 1960s and 1970s there was considerable analytical activity, and during this period ballistic perforation dynamics was examined by Recht and Ipson [7.45]. They analysed the relationship between the energy lost to deformation and heating and the change in kinetic energy of the projectile. Their equations gave good agreement with post-perforation velocities when thin plates were perforated by blunt cylindrical fragments. Perhaps the best documented work during this period was the extensive research carried out in the USA under the guidance of Goldsmith [7.46], whose work has been summarized in a number of fundamental papers. A typical review is contained in reference [7.47], written in the late 1970s.

Much of the fragmentation that causes penetrative damage comes from the explosion of tubular bombs, and it was during the Second World War that the physics of the fragmentation process was first investigated, again by G. I. Taylor [7.48]. He showed analytically that the distribution of stress within the wall of a tube containing detonating explosive was such that longitudinal cracks were likely to form at the outer surface. These cracks penetrate through to the inner wall and cause fragmentation when the internal pressure becomes equal to the tensile strength of the material. The longitudinal cracks in the steel casing of a bomb start close to the detonation wave point, and open out rapidly as the tube expands. Although the cracks are wide, they do not allow the pressurised content to escape until the tube has expanded to double its initial diameter.

The maximum distances that fragments can be driven outwards from explosives depend, of course, on their initial velocities, and these are a function of the size of the explosion. The maximum radial horizontal distance in metres (V) is often given as $45W^{1/3}$, where W is the equivalent weight of TNT charge in kilograms. This was discussed by Kinney and Graham [7.49]. For an ejection angle of α with the horizontal, $r = V^2 \sin^2\alpha/g$, so that the maximum range occurs when $\alpha = 45°$, and at this range the maximum fragment velocity $= (rg)^{1/2}$. Observations of large explosions show that the number and mass of fragments are related exponentially.

During the 1980s much of the analytical work was directed to providing finite element solutions for the action of penetrators on metal targets, and using computer initiated display techniques to plot the successive deformation patterns. The fundamental science was unchanged but the pictorial demonstration was valuable, showing deformed configurations at a given time after impact. It is not the intention here to examine the development of these modelling procedures in detail, but good examples of analytical model development are contained in papers by Ravid and Bodner [7.50], Goldsmith and Finnegan [7.51] and Dehn [7.52]. Much of the work in this field was financed by the military weapon designers, who were seeking improved methods of penetrating armour plating. The plating varied from thick slabs of metal to laminations of metal and composites, and to combinations of metallic and ceramic plating.

At the time of writing, the most recent studies, taking us into the 1990s, include the post-penetration behaviour of the projectile after passing through a metallic plate. This has been discussed in relation to residual velocity and projectile path by Nurick and Crowther [7.53]. It has also been realized that a single theoretical model incorporating all the mechanisms for a given problem, and capable of predicting all the features of the event, has so far proved impossible to develop (to quote Wen and Jones [7.54]). They have reverted to the development of empirical or semi-empirical equations for the perforation of plates struck by blunt or flat-headed missiles, using the principles of dimensional analysis. The kinetic energy of the projectile is taken in two separate elements – one associated with the shearing of a plug of metal having a diameter equal to the projectile, and one representing the energy absorbed by a global structural response. In examining this process, the reader should also consult earlier studies by Neilson [7.55], Jowett [7.56] and Wen and Jones [7.57], which indicate a significant dishing of the plate as well as plug failure. The semi-empirical equations in the work of Wen and Jones are obtained by dimensional analysis.

The equations for the perforation of mild steel plates when struck by blunt or flat-ended projectiles may be summarized (see ref. [7.54]) as:

$$E_p/(\sigma_u\, d^3) = (42.7/10.3)(H/d)^2 + (1/10.3)(s/d)(H/d), \qquad (7.24)$$

where E_p is the perforation energy of the plate (Nm), σ_u is the ultimate tensile strength of the plate (N/m^2), d is the projectile diameter (m), s is the unsupported span of the plate and H is the target thickness (m). This is known as the SRI equation, dating from 1968, originally formulated by Gwaltney [7.58].

A second equation is

$$E_p/(\sigma_u d^3) = 1.4 \times 10^9 (H/d)^{1.5}/\sigma_u, \qquad (7.25)$$

due to Jowett [7.56].

A third equation (see [7.55]) is

$$E_p/(\sigma_u d^3) = 1.4(s/d)^{0.6}(H/d)^{1.7}, \qquad (7.26)$$

and the most recent relationship from Wen and Jones [7.54] is

$$E_p/(K\sigma_a d^3) = (\pi/4)(H/d)^2 + A(s/d)^\alpha (H/d)^\beta, \qquad (7.27)$$

where A, α and β are constants to be determined from experiments, σ_a is an appropriate flow stress and K a constraint factor. Taking $K = 2$ and $\sigma_a = \sigma_y$ (the yield stress) and comparing Eq. (7.27) with experimental results, leads to

$$E_p/(\sigma_u d^3) = (2\sigma_y/\sigma_u)[(\pi/4)(H/d)^2 + (s/d)^{0.21} \cdot (H/d)^{1.47}]. \qquad (7.28)$$

Much of the foregoing work has been connected with missiles versus military armour, missiles versus the protective shields of nuclear reactors, or fragments from exploding jet engines impinging on metal protective shields. A further area of great interest is the effect of missiles and fragments on the thin aluminium sheeting of aircraft structures. A good deal of work in this area has been reported in the publications of the Advisory Group for Aerospace Research and Development (AGARD) of the North Atlantic Treaty Organization (NATO). In an overview of the problem Avery, Porter and Lauzze [7.59] considered three realistic threats: (a) from explosive penetrators, (b) high explosive (HE) projectiles and (c) warheads. They indicated that projectile damage can range from dents, cracks and holes to large petalled areas accompanied by extensive out-of-plane deformation. The measurement that has proved most useful is lateral damage, which is defined as the diameter of a circle that just encloses the limits of fracture, material removal or material deformation (Figure 7.22). Since damage is limited to sheet thickness, projectile speed and impact angle, it is possible to establish a damage regime diagram, as shown in Figure 7.23, which refers to a 0.30 calibre armoured piercing round impacting aluminium sheet of alloy 7075–T6. The variation in damage size with projectile velocity for sheet 0.125 inches thick damaged by 0.30 calibre ball ammunition is shown in Figure 7.24, taken from the work of Avery, Porter and Lauzze. It can be seen that the maximum lateral damage occurs just above the penetration limit, and that further

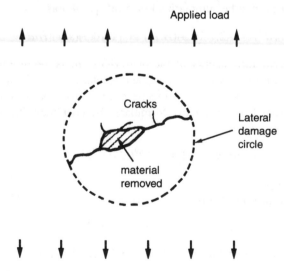

Figure 7.22 Lateral damage circle (from Avery *et al.*, ref. 7.59).

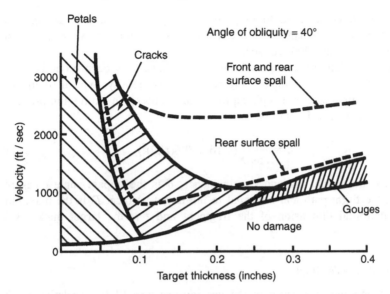

Figure 7.23 Damage type regime diagram for 0.30 calibre armoured piercing round impacting aluminium (from Avery *et al.*, ref. 7.59).

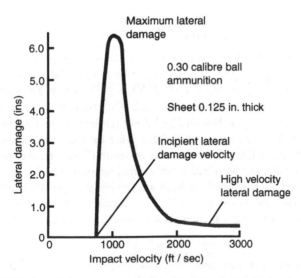

Figure 7.24 Variation of damage size with projectile velocity (from Avery *et al.*, ref. 7.59).

increases in projectile speed do not produce a significant change in damage size. The maximum lateral damage is a circle of radius 6 metres, when the impact velocity is just over 1000 feet/sec.

Damage prediction techniques for fragments have only been properly developed for high-velocity compact fragments of low density, typical of impacts from surface/air missiles or high explosive projectiles. An expression proposed by Avery, Porter and Lauzze is:

$$L = \frac{D}{\cos \theta} (1.16 + 0.6(t/D)^2), \qquad (7.29)$$

where L is the lateral damage size in inches; θ is the impact angle measured between the flight path and a normal to the target surface; D is the maximum projected frontal dimension of the fragment and t is the target thickness in inches.

7.5 FRAGMENTATION

Fragmentation was discussed in the previous section (7.4) in relation to the break-up of the casing of conventional bombs, but the history of fragmentation analysis is an interesting field in its own right. As before, students of the subject are indebted to Christopherson [7.12], who dealt with advances in fragmentation prediction during the Second World War. In 1943 Mott produced three research papers for the UK Armament Research Department which included the theoretical treatment of shell fragmentation (refs [7.60] to [7.62] inclusive). Mott argued that the number of fragments (N) between mass W and $W + dW$ should be given by

$$N = B \, e^{-M/M_A} \cdot dM, \qquad (7.30)$$

where $M = W^{1/2}$ and M_A, B are constants for a given weapon. B was related to the total fragmenting weight, W_0, by the equation $2BM_A^3 = W_0$, and the parameter M_A was given by $M_A = Gt^{5/6}d^{1/3}(1 + t/d)$, where $t =$ thickness of casing (in) and d is the internal diameter (in). Experiments showed that for a TNT filling inside a casing of British Shell Steel (carbon content approximately 0.4 to 0.5%), $G = 0.3$. The fragment weights were measured in ounces.

It was pointed out by Christopherson that equation (7.30) and the formula for M_A were incompatible, but both had experimental support. The limited range of weapons that could be examined by Mott's theory was a greater problem, and this was overcome by Payman [7.63]. His formula related the weight of fragments (W) each of weight greater than w, to W_0 by the equation

$$-cw = \log_{10} W/W_0, \qquad (7.31)$$

where the fragmentation parameter (c) was given by $c = Kt^{-2.45}d^{-0.55}$. The coefficient K depended on the nature of the casing steel and the type of explosive.

In 1943 Payman [7.64] showed that all fragments from an exploding bomb casing containing a relatively feeble explosive, had nearly the same initial velocity, given by:

$$V^2 = 6.95[(E/C)^{1/3} - 0.43]10^7, \tag{7.32}$$

where E and C are the weights per unit length of the charge and the casing respectively in the cylindrical portion of a bomb. This indicated that the velocity would be zero when $E/C< = 0.0795$. Christopherson suggested that this lower limit for E/C should be about 0.2 for less feeble explosives, and proposed the formula:

$$V^2 = 8.22 \times 10^7[1 - e^{-0.69E/C}]. \tag{7.33}$$

Once the fragments have passed through the shock front (because of rapid deceleration of the front) they are travelling in free air. Then they are subject to a retarding force proportional to the square of the velocity. Experiments in the UK in 1943 suggested an equation for velocity, when the velocity was greater than the speed of sound. This was:

$$Vx = Ve^{-0.002xa/W^{1/3}},$$

where W was the fragment mass in ounces, and a was the coefficient of area, given by the equation $a - A_m Q^{-2/3}$. A_m is the mean area presented by the fragment as it rotates during flight, and a is its volume. When the fragment is an exact cube, $a = 1.5$. For velocities below the speed of sound,

$$Vx = Ve^{-0.0014xa/W^{1/3}}. \tag{7.34}$$

The above formulae gave information on risk zones, within which people needed protecting or evacuating.

Christopherson presented calculations to show that the range of fragmentation of a medium case bomb, in which the initial velocity of the fragments would not exceed 7500 ft/sec, was about 3000 ft, which coincided with the generally accepted danger area for a 1000 lb medium capacity bomb. It is interesting to note that 12 or 15 inches of concrete was found to be sufficient to resist the fragmentation of a 500 lb bomb detonated at a range of 50 ft. The corresponding thickness of steel plating was between 1.5 to 1.75 inches.

The fragment velocity equations reviewed by Christopherson, and the simple formula for fragment range (r in metres $= 45W^{1/3}$, where W is kilograms of TNT) were used throughout the period 1950 to 1990, particularly by engineers who were asked to check that an explosion in an ammunition storage unit would not cause a sympathetic explosion in an adjacent unit. In these circumstances, the fragments would be pieces of reinforced concrete, most of which would be propelled at a common velocity. Exploding ammunition stores can also shower human beings with high-velocity fragments, causing injury or death. The Societal risks from this type of disaster have been reviewed by Williams and

Ellinas [7.65] in recent years. Explosive storage regulations in most countries give safe distances between bunkers.

An area of interest has been the combined effect on structures of blast and fragment loading. Observations on the effect of simultaneous loadings have shown that combined loading leads to more severe damage than would be expected if the damage levels from separately occurring effects were added. This is particularly interesting when the spalling of concrete is augmented by blast. Research at the US South West Research Institute in the 1980s indicated that the damage from the combined blast and fragmentation loading of a steel plate was particularly severe. The combined action analysis was discussed by Marchand and Cox in 1989 [7.66].

7.6 DELIVERY SYSTEMS AND THEIR DEVELOPMENT

Projectile and fragment damage to structures is normally the result of military attack, although there are circumstances when a dust, gas or vapour explosion can lead to the fragmentation of the structure causing penetrative damage to an adjacent structure. We will concentrate here on the recent history of weapon development because of its relevance to our general subject. All the information has been drawn from open defence publications and is not therefore of a restricted or confidential nature.

During most of the Second World War the main explosive weapons were bombs, mines, torpedoes and shells, and we have referred to the history of their development in the introduction. A major weapon now is the guided missile in its many forms, with the guidance systems becoming ever more sophisticated.

When a penetrating missile passes through concrete a pressure zone is formed at the interface between the external surface of the penetrator and the surrounding concrete. Material lying in the path of the penetrator must be displaced in a direction transverse to the alignment of the path of entry. This material will crack radially and break into fragments, forming the entry crater surrounded by a cracking zone. When the penetrator strikes the target at an angle very high-pressure differentials can be formed on opposite sides of the penetrating point, which can deform and deflect the penetrator. This is a complex situation to analyse, so a considerable financial commitment has been made by many countries to sophisticated experimental facilities.

Firing tests against concrete targets with conventional bombs or gas guns for weapons up to about 100 kg in weight have been well documented, but for larger penetrators (above 100 kg and perhaps as much as 1000 kg) countries such as the USA and France use rocket sleds, and in Germany a light-gas acceleration gun is used. The latter equipment can decelerate a 1000 kg mass to 300 m/s within 29 m.

Hardened targets need to be attacked by warheads that penetrate before exploding, so the penetrative equipment must be guided with great precision, hence the need to produce weapons such as laser guided bombs. Experience in the Second World War, Vietnam and elsewhere have emphasized the poor performance of conventional bombing against targets such as bridges. Many hundreds of sorties against a bridge, using conventional bombs, could still fail to destroy the target. The solution is to add 'guided bomb units' to conventional weapons. During the Gulf War conventional bombs could not properly engage Iraqi bunkers, often set as far as 30 m below ground with reinforced concrete roofs many metres thick. The US Air Force therefore designed a bomb case with a penetrator manufactured from a 203 mm diameter gun barrel, giving a total weight of 2132 kg and carrying 295 kg of Tritonal explosive. Apparently this weapon could penetrate 6.5 m of steel-reinforced concrete.

Data from the US suggests that a guided bomb is over 100 times as cost effective against point targets as an unguided bomb, and without precision bombs it would not have been possible to destroy 54 main bridges over the Euphrates and the Tigris during the Gulf War. Further discussion of these points is contained in a review by Niemzig and Steffen [7.67], who indicate that the cost of using a guided bomb is about the same as using a cruise missile. Precision bombing systems need enhanced accuracy, and this was first made possible by the introduction of digital computing techniques based on solid state hardware. High accuracy navigation data could be provided by systems based on inertial navigation sensors, which are basically a combination of gyroscopes and accelerometers controlled by a microprocessor. This configuration provides aircraft position, velocity and other information from which navigation and steering data can be computed. From this data, in conjunction with weapon ballistic information, release and pull-up points can be calculated. In the mid-1980s a system was introduced based on a miniature gimballed inertial platform used in conjunction with a laser range finder. This produced accurate navigation, weapon control and a means of precision designation.

As well as the kinetic energy penetrator, which is a solid component projected at high velocity by high-pressure guns and which relies on kinetic energy to pierce the target, the shaped charge is also used to pierce armour. This well-known type of charge consists of a copper-lined cone embedded in an explosive cylinder. When a fast fuse detonates the high explosive charge, the detonation wave squeezes the liner into the form of a long, high-speed jet, which generates very high pressures on a steel target when contact takes place. These pressures are an order of magnitude greater than the yield strength of armour plate, forcing the material to flow in the hydrodynamic manner of soil or mud under the influence of a strong water jet. The effectiveness of the shaped charge is independent of the speed of delivery of the projectile, and because of this it is suitable for missiles and rocket launchers with relatively low missile velocity.

Although we are only concerned with the structural loading and penetrative action in this chapter, it is interesting to look quickly at the analysis of shaped charge action and at methods to combat this effectively. The analysis was established by G. I. Taylor and Schardin (ref. [7.68, 7.69]) who showed that the maximum penetration and hole size can be calculated by the conservation of energy and momentum. The piercing power is proportional to the length of the jet and the square root of its density.

A shape charge performs at its best when the distance from the face of a target to the point of detonation is precisely calculated (the stand-off). For this reason many shaped charge missile warheads have an extended stand-off probe to initiate detonation. Manufacturers of shaped charge warheads are now seeking penetrative powers equal to six times the warhead diameter. However, there is a problem because warheads with long probes tend to become unstable in flight.

The shaped charge liner is normally made of copper, and technology advances have usually been improvements in the precision of the liner shaping and centring. Efforts have also been made to produce detonation wave fronts normal to the cone surface, so that the liner remains solid but changes shape like a fluid. The high-velocity jet formed by the transformation of the metal liner tends to break into separate particles, but the development of the explosively formed projectile, or self-forming fragment, has meant the retention of full effectiveness over long distances.

If the liner is made parabolic in shape, rather than conical, the slug, instead of being needle shaped, becomes a chunky, U-shaped projectile. This type of projectile does not burst into fragments or flatten against the target, but the penetrating power is naturally reduced.

In a shaped charge, the initial velocity of the jet depends on the ratio (μ) between the mass of the liner and the mass of explosive located behind the liner element. This velocity, V_1, is given by $V_1 = A(\mu + k)^{-1/2}$, where A is a constant defining the specific energy of the explosive, and k is a charge configuration constant. This formula was due to Gurney [7.70]. A less approximate analysis of the relationship between V_1 and the charge configuration has been given more recently by Chou *et al.* [7.71].

The defeat of shaped charge jets by armour material was a problem that absorbed much military research and development in the years following the Second World War. Eventually these studies turned towards the idea of 'active' or energetic armour elements, which would release energy when attacked that would counteract the effect of the charge. The principle of operation was taken an important step forward in the 1970s, by the invention of the 'drive-plate' explosive sandwich concertina of two metal plates with an explosive layer between. When initiated by a high-speed jet, the explosive drives the metal plates to one side, and the moving plates absorb the mass of the jet. Work began in Israel in 1974 on prototype 'add-on' kits for tanks, which would cover the turret and the forward part of the hull. It was important to show that the explosive layer in the sandwich would not be initiated by any weapon other than the high-velocity jet of a shaped charge. The kits finally took the form of 'blocks' or 'tiles' that could be fixed to the external surfaces, and which, in total, did not add more than 100 kg to the weight of the tank. Reactive armour first appeared on the battlefield in 1982, and one of the first unclassified accounts of its development history was given by Vered [7.72] in 1987.

7.7 REFERENCES

7.1 Poncelet, J. V. (1829 and 1835) *Cours de mécanique industrielle*, Paris.

7.2 Timoshenko, S. (1953) *History of Strength of Materials*, McGraw-Hill.

7.3 Petry (1910) *Monographies de systèmes d'artillerie*, Brussels.

7.4 Young, C. W. (1972) *Empirical Equations for Predicting Penetration Performance in Layered Earth Materials for Complex Penetrator Configurations*, Report SC-DR-72-0523, Sandia Laboratories, Albuquerque, December.

7.5 Goodier, J. N. (1965) On the mechanics of indentation and cratering in solic targets of strain-hardening metal by impact of hard and soft spheres, *Proc. 7th Symposium on Hypervelocity Impact*, Vol. 3, p. 215.

7.6 Norwood, F. R. (1974) *Cylindrical Cavity Expansion in a Locking Soil*, Report SLA-74-0261, Sandia Laboratories, Albuquerque, July.

7.7 Yew, C. H. and Stirbis, P. P. (1978) Penetration of projectile into terrestrial target, *Journal of the Engineering Mechanics Division*, ASCE, EM2, paper 13703, April, p. 273.

7.8 Allen, W. A., Mayfield, E. B. and Morrison, H. L. (1957) Dynamics of a projectile penetrating sand, *Journal of Applied Physics*, **28**(3), March.

7.9 Robins, B. (1742) *New Principles of Gunnery*, London.

7.10 Euler, L. (18th century) Neue Grundsätze der Artillerie, Berlin.

7.11 Stipe, J. G. (1946) *Terminal Ballistics of Soil, Effects of Impacts and Explosions*, Summary Technical Report of Division 2, NDRC, Vol. 1, Washington.

7.12 Christopherson, D. G. (1946) *Structural Defence* (1945), UK Ministry of Home Security, Civil Defence Research Committee paper RC 450.

7.13 US Army (1965) *Fundamentals of Protective Design (Non-nuclear)*, US Army TM5-855-1, Department of the US Army, July.

7.14 Young, C. W. and Keck, L. J. (1971) *An Air-dropped Sea-ice Penetrometer*, Report SC-DR-71-0729, Sandia Laboratories, Albuquerque, New Mexico, USA, December.

7.15 Forrestal, M. J., Longcope, D. B. and Lee, L. M. (1983) Analytical and experimental studies on penetration into geological targets, *Symp. on Interaction of Non-nuclear Munitions with Structures*, US Air Force Academy, Colorado, May.

7.16 Bishop, R. F., Hill, R. and Mott, N. F. (1945) The theory of indentation and hardness tests, *Proc. of the Physical Society*, **57**(3), May.

7.17 Young, C. W. (1969) Depth prediction for earth penetrating projectile, *Journal of Soil Mechanics and Foundation Division*, ASCE, May.

7.18 Stilp, A., Schneider, E. and Hülsewig, M. (1985) Penetration of fragments into sand, *Second Symp. on Interaction of Non-nuclear Munitions with Structures*, Panama City Beach, Florida, April, p. 333.

7.19 Schoof, L. A., Maestas, F. A. and Young, C. W. (1989) Numerical method to predict projectile penetration, *Fourth International Symposium on Interaction of Non-nuclear Munitions with Structures*, Panama City Beach, Florida, April.

7.20 Taylor, T. and Fragaszy, R. J. (1989) Implications of Centrifuge Penetration Tests for use of Young's equation, *Fourth International Symposium on Interaction of Non-nuclear Munitions with Structures*, Panama City Beach, Florida, April.

7.21 Gebara, J. M., Pau, A. D. and Anderson, J. B. (1993) Analysis of rock-rubble overlay protection of structures by the finite block method, *Proc. of Sixth International Symp. on Interaction of Non-nuclear Munitions with Structures*, Panama City Beach, Florida, May, p. 12.

7.22 Nelson, R. B., Ito, Y. M., Burks, D. E. *et al.* (1983) *Numerical Analysis of Projectile Penetration into Boulder Screens*, US Army Engineer Waterways Experiment Station Report WES.MP.SL.83.11.

7.23 Gelman, M. D., Richard, B. N. and Ito, Y. M. (1987) *Impact of AP Projectile into Array of Large Caliber Boulders*, US Army Engineer Waterways Experiment Station.

7.24 Chen, W. F. and Pau, A. D. (1991) Finite element and Finite block methods in Geomechanics, *Proc. of Third Int. Conf. on Constitutive Laws of Engineering Materials*, Tucson, Arizona.

7.25 Schwer, L. E., Rosinsky, R. and Day, J. A. (1987) Lagrangian technique for predicting earth penetration including penetrator response, *Third International Symposium on Interaction of Non-nuclear Munitions with Structures*, Mannheim, Germany, March.

7.26 Creighton, D. C. (1982) *Non-normal Projectile Penetration in Soil and Rock: User's Guide for Computer Code PENCO2D*, US Army Waterways Experiment Station Technical Report SL.82.7, Vicksburg, September.

7.27 Matuska, D., Durrett, R. E. and Osborn, J. J. (1982) *HULL User's Guide for Three Dimensional Linking with EPIC3*, Orlando Technology Inc. Report ARBRL-CR-00484, Shalimar, Florida, July.

7.28 Ito, Y. M., Nelson, R. B. and Ross-Perry, F. W. (1979) *Three Dimensional Numerical Analysis of Earth Penetration Dynamics*, US Defense Nuclear Agency Report DNA 5404 F, January.

7.29 Yankelevsky, D. Z. (1983) Projectile penetration through a narrow drill in soil, *Int. Journal of Impact Engineering*, 1(4), p. 377.

7.30 Young, C. W. and Young, E. R. (1985) *Simplified Analytical Model of Penetration with Lateral Loading*, Sandia National Laboratories Report SAND-84-1635, May.

7.31 Amini, A. and Anderson, J. B. (1993) Modelling of projectile penetration into geologic targets based on energy tracking and momentum impulse principle, *Proc. of Sixth International Symp. on Interaction of Non-nuclear Munitions with Structures*, Panama City Beach, Florida, May, p. 6.

7.32 Pariseau, W. G. and Fairhurst, C. (1967) The force penetration characteristic for wedge penetration into rock, *Int. Journal of Rock Mech. and Min. Science*, **4**, p. 165.

7.33 Anderson, W. F., Watson, A. J., Johnson, M. R. and McNeil, G. M. (1983) Optimization of rock/polymer composites to resist projectile penetration, *Materiaux et Constructions*, **16**(95), 343.

7.34 Whiffen, P. (1943) *UK Road Research Laboratory Note No. MOS/311*.

7.35 Beth, R. A. (1946) Effects of impact and explosion, *Terminal Ballistics of Concrete*, Chap. 7, Vol. 1, Summary Tech Rep DW2, NRDC, Washington.

7.36 Forrestal, M. J., Altman, B. S., Cargile, J. D. and Hanchak, S. J. (1993) An empirical equation for penetration depth of ogive-nosed projectiles into concrete targets, *Proc. of 6th Int. Symp. on Interaction of Non-nuclear Munitions with Structures*, Panama City Beach, Florida, USA, May.

7.37 Brown, S. J. (1986) Energy release protection for pressurised systems, Part 2. Review of studies into impact/terminal ballistics, *Applied Mechanics Review*, **39**.

7.38 Gold, V. M., Pearson, J. C. and Turci, J. (1993) Resistance of concrete and reinforced concrete structures impacted by CU and TA projectiles, *Proc. of 6th Symp. on the Interaction of Non-nuclear Munitions with Structures*, Panama City Beach, Florida USA, May.

7.39 Tate, A. (no date) A theory for the deceleration of long rods after impact, *J. Mech. Phys. Solids*, **15**, 387.

7.40 Peters, J., Anderson, W. F. and Watson, A. J. (1994) Determination of penetration parameters from the stress pulse generated during projectile penetration into cementitious materials, *Structures Under Shock and Impact III*, Comp. Mech. Pub., June, p. 235.

7.41 Dancygier, A. N. and Yankelevsky, D. Z. (1994) Comparative penetration tests of fibre reinforced concrete plates, *Structures under Shock and Impact III*, Comp. Mech. Pub., June, p. 273.

7.42 Taylor, G. I. (1948) The formation and enlargement of a circular hole in a thin plastic sheet, *Quarterly Journal of Mechanics and Applied Maths*, **1**, p. 103.
7.43 Taylor, G. I. (no date) The use of flat ended projectiles for determining dynamic yield stress, *Proc. Roy. Soc. London*, Vol. 194, Series A, p. 289.
7.44 Thomson, W. T. (1955) An approximate theory of armour penetration, *Journal of Applied Physics*, **26**(1), January.
7.45 Recht, R. F. and Ipson, T. W. (1963) Ballistic perforation dynamics, *Journal of Applied Mechanics*, September.
7.46 Goldsmith, W. *et al.* (1965) Plate impact and perforation by projectiles, *Experimental Mechanics*, December.
7.47 Backman, M. E. and Goldsmith, W. (1978) The mechanism of penetration of projectiles into targets, *Inter. Jour. Eng. Sci.*, **16**, Pergamon Press, London.
7.48 Taylor, G. I. (1944) *The Fragmentation of Tubular Bombs*, Paper written for the Advisory Council on Scientific Research and Technical Development, UK Ministry of Supply.
7.49 Kinney, G. and Graham, K. J. (1985) *Explosive Shocks in Air*, Springer Verlag, New York.
7.50 Ravid, M. and Bodner, S. R. (1983) Dynamic perforation of viseoelastic plates by rigid projectiles, *Int. Eng. Sci.*, **21**, 577.
7.51 Goldsmith, W. and Finnegan, S. A. (1986) Normal and oblique impact of cylinders – conical and cylindrical projectiles on metallic plates, *Int. J. Impact Engng*, **4**, p. 83.
7.52 Dehn, J. (1989) Mathematical models in penetration mechanics, *Structures Under Shock and Impact* (1), Elsevier/C M Publications, p. 203.
7.53 Nurick, G. N. and Crowther, B. G. (1994) The measurement of the residual velocity and deflection of a projectile passing through a thin plate, *Structures Under Shock and Impact* (3), CM Publications, p. 261.
7.54 Wen, H. M. and Jones, N. (1992) Semi-empirical equations for the perforation of plates struck by a mass, *Structures Under Shock and Impact* (2), Thomas Telford and CM Publications, p. 369.
7.55 Neilson, A. S. (1985) Empirical equations for the perforation of mild steel plates, *Int. J. Impact Eng.*, **3**, p. 137.
7.56 Jowett, J. (1986) *The Effect of Missile Impact on Thin Metal Structures*, UKAEA SRD R378.
7.57 Wen, H. M. and Jones, N. (1992) *Experimental Investigation into the Dynamic Plastic Response and Perforation of a Clamped Circular Plate Struck Transversely by a Mass*, Impact Res Centre ES/85/92, Dept of Mechanical Engineering, University of Liverpool.
7.58 Gwaltney, R. C. (1968) *Missile Generation and Protection in Light Water-cooled Power Reactor Plants*, Report URNL-USTC-22, Oak Ridge National Lab., Tennessee, USA.
7.59 Avery, J. G., Porter, T. R. and Lauzze, R. W. (1975) Structural integrity requirements for projectile impact damage – an overview, Specialist meeting on Impact Damage Tolerance of Structures, *AGARD Conf. Proceedings*, No. 186.
7.60 Mott, N. F. (1943) *Fragmentation of H E Shells: a Theoretical Formula for the Distribution of Weights of Fragments*, UK Armament Research Department, AC 3642 SD/FP85, March.
7.61 Mott, N. F. (1943) *A Theory of Fragmentation of Shells and Bombs*, UK Armament Research Department, AC 4035, SD/FP 106, May.
7.62 Mott, N. F. (1943) *Fragmentation of Shell Casings and the Theory of Rupture in Metals*, UK Armament Research Department, AC 4613 SD/FP 139, August.
7.63 Payman, W. (1943) *Empirical Relationships for Use in Comparing Results of*

Fragmentation Trials, I – Velocities of Fragments, UK Safety in Mines Research Station, RC 387, June.

7.64 Payman, M. (1943) *Empirical Relationships for Use in Comparing Results of Fragmentation Trials, II – Weight Distribution of Fragments*, UK Safety in Mines Research Station, RC 388, July.

7.65 Williams, K. A. J. and Ellinas, C. P. (1989) Societal risks from detonations in explosive storage structures, *Structures Under Shock and Impact*, Elsevier and CM Publications, p. 253.

7.66 Marchand, K. A. and Cox, P. A. (1989) The apparent synergism in combined blast and fragment loadings, *Structures Under Shock and Impact*, Elsevier and CM Publications, p. 239.

7.67 Niemzig, I. and Steffen, R. (1992) Engagement of hard targets from the air, *Military Technology*, November, p. 48.

7.68 Taylor, G. I. (1941) *The Analysis of a Long Cylindrical Bomb Detonated at One End*, Paper written for the Civil Defence Research Committee, UK Ministry of Home Security.

7.69 Schardin, H. (1956) Uber das Wesen der Hohlladung, *VDI Journal*, **98**, 1837.

7.70 Gurney, R. W. (1943) *The Initial Velocity of Fragments from Bombs, Shells and Grenades*, US Ballistic Research Laboratory, BRL – Report 405.

7.71 Chou, P. C. *et al.* (1981) Improved formula for velocity, acceleration and projection angle of explosively driven liners, *Proc. 6th Int. Symp. Ballistics*, Orlando.

7.72 Vered, G. R. (1987) Evolution of 'Blazer' reactive armour and its adaptation to AFVs, *Military Technology*, December, p. 53.

8

The effects of explosive loading

8.1 CIVIL BUILDINGS

There have been so many terrorist attacks in recent years on commercial, government and private buildings that a considerable bank of information exists on structural behaviour in the face of man made explosions. There have been spectacular gas and vapour cloud explosions in many countries of the world, all well documented and the subject of subsequent structural investigation. But a number of classic investigations of the effect of aerial bombing on civil buildings were made during the Second World War, and it is sensible to take as a starting point the papers by Thomas [8.1] and Baker, Leader Williams and Lax [8.2] that appeared in the *Civil Engineers in War*, published by the UK Institution of Civil Engineers in 1948. The diagrams and photographs look remarkably similar to those accompanying articles written in 1994 on the effect of terrorist bombing on the City of London.

Baker *et al.* noted that fully framed concrete or steel buildings were very resistant to the effect of bombing during the Second World War. Small and medium-sized bombs dropped during that conflict were found to damage framed buildings by direct hits, but the blast from near misses of similar bombs caused only superficial (though extensive) damage. Direct hits often create excessive internal pressure within buildings, which reverses the loads for which the structure was originally designed. Very much larger bombs produce very serious external blast loads that cause downwards failure of roof trusses and lateral distortion of the entire building.

The authors made a very important survey of about fifty cases of damage by high explosive bombs to multi-storey steel-framed buildings. The general Second World War experience was that the limit of vertical bomb penetration was the roof and two or three floors, but bombs released from low altitude could attack lower floors by passing through openings or penetrating external walls. The cutting of a single beam or stanchion did not usually result in collapse of the complete structure, but if a medium-sized bomb exploded inside a building it was found that about 1000 square feet of floor was demolished immediately above and below the explosion. The area of demolition fell off rapidly at lower floors, soon down to 100 square feet. Debris was usually held by the second and third floor slabs below the explosion, but occasionally it brought down a succession of floors.

The survey emphasized that the virtual absence of progressive collapse in well-designed frame buildings meant that floors and walls near the detonation point were effective as screens. Of equal importance was the ductility and continuity of the framework, which made the building highly resistant to explosions. A possible weakness, however, was the lack of ductility and continuity in bolted or riveted connections. As in many other examples of structural loading, the crucial factor lay in the strength or weakness of connections. During the years before the Second World War welded connections were relatively rare, so there was not a great deal of evidence about the behaviour of welded frameworks. The authors, however, drew attention to the possibility of introducing ductility into building structures by clamped connections that rely on friction to produce the effects of continuity. The resulting structure could absorb a large amount of energy without collapse.

The following types of damage were noted in single-storey steel-framed buildings such as hangars, warehouses and workshops: direct damage, primary collapse and spreading collapse. Direct damage was the cutting of members by a bomb (before exploding), by fragments or by crater debris; primary collapse was the collapse of members that depended for their stability on a destroyed member. Spreading collapse occurred when the forces set up by a primary collapse were transmitted to adjoining undamaged members, producing instability in these members. The spreading collapse of a roof could occur when the cutting of a single member suddenly involved the whole area of a large structure. The chief risk to roof steelwork arose from the failure of stanchion to roof-girder connections, which were sheared by violent displacement or lateral blast, or destroyed by blast uplift.

Dwelling houses, of course, are rarely constructed with a ductile steel framework, so in assessing the behaviour of houses in explosions, it was necessary to examine the strength of masonry and brick structures to blast.

Civilian masonry structures are not normally designed to resist explosive blast. The stone castles and military fortifications of the Middle Ages were more likely to be threatened by the penetrative action of missiles, and their wall thicknesses were set empirically by the need to resist damage by local impact and fragmentation. Civil masonry and brick structures were known to have relatively little resistance to local explosions, and in earlier times no attempt was made to predict how they might behave under attack, or what their residual strength might be. Masonry and stone were not employed much in the construction of 'bomb-proof' shelters once reinforced concrete appeared on the structural scene. As the Second World War approached, however, experimental research was initiated to check the behaviour of conventional building structures when subjected to the general blast from exploding aerial bombs, and by 1940 there were several UK reports on the behaviour of brick wall panels.

Most domestic brick walls were 9 in thick, and only a small number of factory or civic buildings were constructed with 13.5 in thick walls. Christopherson [8.3] reported that the static pressure to cause collapse of an 8.5 ft square panel of 9 in brickwork was about 2.25 lb/in², and that the ultimate deflection for walls enclosed in steel channels was about 5 in. For 13.5 in walls the figures were 5.3 lb/in² and 8 in respectively. In terms of hydrostatic impulse, it was shown that for a very wide range of incident pressures the impulse needed to destroy 9 in brickwork was 92 ± 10 lb msec/in². A typical two-storey brick house of the 1940s had free wall spans between 8 and 10 feet square, and field observations following air raids on British cities showed that damage involving the collapse of at least one external wall occurred when the hydrostatic impulse lay between 64 and 84 lb msec/in². Complete demolition of more than 75% of all external walls occurred at impulses between 82 and 160 lb msec/in².

During the Second World War it was necessary to use protective brick walls to reduce blast damage on industrial constructions, and in the early years of the war the design criterion was that collapse of a wall could be avoided if

$$\rho_0 M w_u > 0.25 A^2, \tag{8.1}$$

where ρ_0 is the static yield pressure, w_u is the maximum permissible displacement, M is virtual mass ($= 0.55 \times$ true mass), and A the positive impulse of the explosion. There is an inherent safety margin in this analysis, because it assumes that there is no reduction in impulse by diffraction, and that the brick wall is rigidly supported. In practice diffraction can reduce impulse and the wall could be free to slide as a unit, underlining the importance of ductility in defensive construction. Values of impulse needed to destroy thick panels, taken from ref. [8.3] are shown in Figure 8.1.

The advent of nuclear explosions and the need to assess the behaviour of civilian housing under the long duration blast associated with the detonation of a nuclear bomb led to much field research in the late 1940s and early 1950s. Brick houses were built at the nuclear test sites and destroyed by the blast winds. It was generally assumed that a peak nuclear overpressure of 5 lb/in² would be sufficient to cause the collapse of a brick structure by implosion. For conventional weapons, with a much reduced positive phase duration, higher initial overpressures can be tolerated. It has been reported by Eytan [8.4] that observed damage to Israeli structures during the period 1968–91, where the damage was inflicted by car bombs, explosive charges, artillery shells, air bombs and missiles, showed that normal cavity brick walls would be severely damaged by 7 bars overpressure and would collapse under 12 bars.

Locally damaged brick walls in buildings often transmit loads due to structural weight by an arching action, as indicated by Rhodes [8.5] and by Williams *et al.* [8.6]. These authors also note that the capacity of brickwork or masonry to lean or bulge without collapse is largely governed by the quality of the mortar between individual bricks or masonry blocks. Here the age of the structure is important, because older structures have mortar with a low cement content and low adhesion.

During the early years of the Second World War the British built surface shelters of reinforced brick to protect the civilian population from the blast and fragmentation resulting from the detonation of a 500 lb bomb filled with TNT at a distance of 50 feet. Later this distance was reduced to 15 feet. It was also thought necessary to provide protection against the earth shock from a delay-fused bomb, because a brick shelter could disintegrate under earth shock effects or move so rapidly that severe impact occurred when it struck the ground after projection through the air. Tests showed that unreinforced brick surface shelters broke up under the action of a 500 lb bomb buried 12.5 ft below ground at a distance of 15 ft horizontally from the shelter wall. Roof, floor and walls separated, or the cross section folded up like an elongated parallelogram. Break-up was prevented by reinforcing the brickwork with a small percentage of steel bars (as little as 0.06% by volume) carried continuously from the floor through the walls and into the roof. Folding-up was prevented by introducing cross walls at intervals along the shelter length. Sometimes the exterior of the brick shelter was covered by steel mesh, held in position by an additional 4.5 in brick skin. A

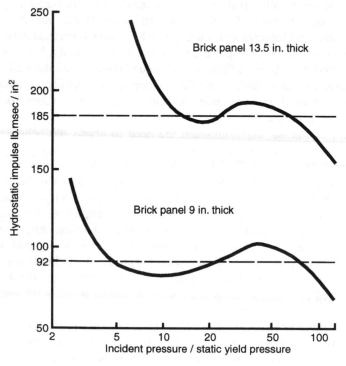

Figure 8.1 Impulses for the destruction of brick panels (from Christopherson, ref. 8.3).

typical reinforced brick shelter used brick side walls combined with a concrete floor and roof. Reinforcing bars were introduced into the cavity between bricks before the cavity was filled with concrete. When attacked by blast weapons surface brick shelters were of course much more vulnerable than buried shelters.

Brickwork was also protected during the Second World War by internal strutting, which was placed so that there was enough lateral stability to eliminate the possibility of collapse from unsymmetrical loading. Strutting was also used to support the load of debris on the roof of the building if destruction of the walls occurred. Tests on brick shelters during the Second World War were described in a series of reports by the Research and Experiment department, Ministry of Home Security, and examples are given in refs [8.7] to [8.9]. The lateral velocity imparted to a surface shelter by earthshock from a bomb of 250 kg at 15 ft horizontally from the shelter wall was found to be about 10 ft/sec, as measured and reported by Walley [8.10].

The amount of reinforcement required to keep a brick surface shelter intact was calculated very simply by Christopherson. He supposed that when an unreinforced shelter breaks up the maximum relative velocity of roof and walls is 5 ft/sec for a shelter 30 ft × 8 ft in plan area with walls 1 ft thick. If the roof is assumed to be 5 in thick, with a density of 144 lb/ft^3, its kinetic energy is $(60 \times 25)/64$ ft lb/ft^2. The kinetic energy of the roof per foot run of wall is $(60 \times 25 \times 240)/64 \times 72$ ft lb. If there are A in^2 of reinforcing bars per ft run of wall, having a yield stress of 50 000 lb/in^2, and if their extension is d ft, then $A.d.$ 50 000 = the kinetic energy per foot run. If, as suggested earlier, a minimum area of reinforcement of 0.06% is used, then in a 1 ft thick wall, the value of $A = 0.06 \times 1.44$, and $d = (60 \times 25 \times 240)/(64 \times 72 \times 50\,000 \times 0.06 \times 1.44) = 0.22$ in. Thus the separation of wall and roof cannot be greater than about $\frac{1}{4}$ in, and could well be less. This indicated that the shelter components would remain virtually intact.

This wartime research and testing was still used as a basis for design rules over the next thirty or forty years. Thus, when the US Army Technical Manual on Protective Design (Non-nuclear) appeared in 1965 [8.11], it contained wartime information on the thickness of reinforced and plain brickwork to provide protection against fragments and blast from general purpose bombs detonated at a distance of 40 feet. The required thicknesses are shown in Table 8.1.

It was also recommended that brick structures designed with these thicknesses should be spaced 120 ft apart. Blast resistant protective brick walls were also required to be buttressed at intervals not greater than 10 ft, with the buttresses reinforced horizontally.

In Britain, the response of brick walls was often based on information given by a wartime bulletin issued by the Ministry of Home Security [8.12], which gave the minimum distance in feet at which only slight damage would occur from surface blast. The required distances are shown in Table 8.2.

Table 8.1

	Wall thickness in inches	
Bomb size (lb)	Reinforced brickwork	Plain brickwork
100	13.5	13.5
250	13.5	17.0
500	17.0	21.5
1000	21.5	25.0
2000	25.0	28.5

It is worth noting that, according to this experimentally based information, the safe distances were directly proportional to thickness, and increases in distance were very roughly proportional to the square root of bomb size increases. It was generally assumed that for conventional blast the period of vibration of a brick structure would be at least 10 times greater than the duration of the pressure pulse, and that structural response could be calculated without the need to resort to single degree of freedom methods. However, for a nuclear blast the fundamental period of the structure could be less than the blast duration, and more complex methods of establishing structural response were necessary. Ultimate strength calculations for brick panels were based on yield line theory for slabs, using an ultimate stress in bending of 1.5 N/mm^2 and a modulus of Elasticity of 7000 N/mm^2.

Most of the information given above can be roughly assumed to apply to edge supported brick walls measuring 8 ft square, but in the case of nuclear blast the destruction of brickwork is usually related to the complete destruction of brick houses. The results of nuclear tests in the early 1960s suggested that for a conventional brick dwelling house a nuclear explosion would cause serious damage at a distance equivalent to a peak overpressure of about 3.5 psi. Complete destruction would occur at a peak overpressure of 5 psi.

Table 8.2

	Minimum distance in feet		
Bomb size (kg)	4.5 in wall	9 in wall	13.5 in wall
50	50	25	17
250	150	80	50
500	240	120	80
1000	360	180	120
1800 (GP)	520	260	170
1800 (light case)	750	370	150

The stability of masonry walls under the action of blast from conventional explosives was examined experimentally in a series of tests in the late 1970s and early 1980s in the USA. Collapse mechanisms were recorded, as well as the nature and distribution of debris from shattering walls, and a range of construction was investigated, varying from load-bearing masonry to reinforced concrete frames with a masonry infill. These tests were summarised by Rempel and Beck [8.13], who also analysed crushing energy and wall rotation.

Attention was also paid to the design of explosive storage facilities constructed of masonry, including overpressures to cause incipient failure and safe separation distances between storage facilities and neighbouring buildings of brick, masonry and prefabricated blocks. The state of affairs in the early 1980s was reviewed by Napadensky and Longinow [8.14]. They quoted current US regulations, which at the time permitted conventional, inhabited buildings to be located at distances of 40 $W^{1/3}$ to 50 $W^{1/3}$ ft from an explosives store, where W is the weight of explosive in pounds. For long-duration pressure/duration blast loads that would be experienced in a nuclear explosion, incipient failure overpressures (where the structure will fail under any additional load) was given in terms of probability of failure as follows.

(a) Light commercial buildings with masonry load bearing walls: 2.0; 3.2; 4.9 psi for failure probabilities of 10%, 50% and 90% respectively.

(b) One-storey masonry load-bearing wall: 1.8; 2.8; 4.6 psi for failure probabilities of 10%, 50% and 90% respectively.

Safe separation distances, peak overpressures and impulses, for a range of non-nuclear explosions, were tabulated by the authors for a range of targets. Excerpts from their survey are given in Table 8.3.

The authors compared the damage, pressures and distances with the recorded effects of the Flixborough vapour cloud explosion in 1974. Damage analysis

Table 8.3

Explosive quantity (lb):	Safe separation distances (R), peak overpressures (P), and impulse (I)											
	1000			10 000			100 000			1 000 000		
	R	P	I	R	P	I	R	P	I	R	P	I
House (split level), frame-brick	290	1.79	30	776	1.36	54	2089	1.0	92	4500	1.0	190
Office building, multi-storey, block walls	140	5.68	61	323	4.6	128	928	3.0	205	2500	2.15	340
School, one-storey, masonry	180	3.64	48	496	2.49	84	1671	1.36	116	5500	0.77	155

(Separation distance in ft; overpressive in psi; impulse in psi ms)

suggested that the blast effects were similar to those produced by 35 300 lb of TNT detonated at a height of 147.6 ft (45 m). According to US safety standards, brick buildings farther away than 1312 ft (400 m) would have been safe, but the damage extended beyond this distance. It was concluded that for this type of vapour cloud explosion damage is directly related to peak overpressure.

Gas explosions in dwelling houses can cause the outward deformation of brick, masonry or precast wall panels. Structural loads due to a range of gas explosions have been measured by the UK Building Research Establishment, where pressures and wall displacement time histories were recorded during the demolition of maisonette blocks. The results have been summarized by Ellis and Crowhurst in two reports ([8.15] and [8.16]). Aerosol canisters containing butane were ruptured and the contents ignited to produce internal peak pressures similar to those in domestic gas explosions. Peak pressures in the range 2.6 KN/m^2 to 9.0 KN/m^2 were generated, depending on the size of the canister. Although structural damage to partitions and internal doors was significant, there was no permanent damage to main structural walls. The largest measured displacement of a wall panel was 0.63 mm for a 9.0 KN/m^2 explosion (750 ml canister). Typical room volumes ranged from 17.5 m^3 (smaller bedrooms and kitchens) to 28.1 m^3 (large second-floor bedroom).

A collation of reports of gas explosion damage to domestic buildings has been made by Moore [8.17] in which explosions are classed in terms of the severity of damage, as moderate, severe and very severe. A moderate explosion can destroy a relatively weak structure, such as a small brick or masonry bungalow.

The effect of blast on dwelling houses of brick was also investigated during the Second World War by the US authorities. Experiments at full scale were undertaken at the Aberdeen Proving Grounds of the Corps of Engineers, and at Princeton University, and the results were summarized in the NRDC technical report on the effects of impact and explosion published in 1946 [8.18]. One series of tests concerned a structure, 6.5 ft square, 4 ft high, with a reinforced concrete floor, roof and columns, enclosed by brick walls, which were bonded to the framework along their bottom and sides but which were free at the top. Charges consisting of 22 and 44 grams of tetryl explosive were detonated at the centre of the enclosed space, and 0.5 and 1.0 lb charges of TNT were exploded outside the structure, 3 ft from the centre of the brick wall.

The 44 gram charge completely blew out one wall (one course thick), whereas the 22 gram charge caused enough cracking to reduce structural strength to zero, without actually blowing the wall outwards. The external detonation of 0.5 and 1.0 lb charges of TNT 3 feet from the centres of already cracked brick walls showed no appreciable additional effect. When a 0.5 lb charge was detonated within the structure complete destruction resulted.

At larger scales (20 ft square, 12 ft high), with 12 in thick brick walls, it was calculated that internal explosions of 1.25 lb TNT would cause serious wall cracking, that 2.5 lb would blow out the walls without destroying the frame, and 15 lb would cause complete destruction. 30 lb of TNT detonated externally a distance of 10 ft from the wall would not seriously damage it. It should be noted that these early experiments were made with no venting to reduce the degree of confinement.

Glass and light sheeting in buildings are very damage-prone under the action of blast loads or penetrators. Unless treated on the surface by the addition of film or mesh, glass will rapidly shatter and be transformed into high-speed splintered fragments. Light sheeting will tear and distort if made from ductile materials, or shatter like glass if made from brittle materials. The loading will be not unlike that calculated for similar areas of brickwork, but the effects of the blast will be felt over larger areas.

Damage from terrorist bombs in the commercial areas of cities can be widespread and expensive to repair. Readers will have seen photographs of the distortion and tearing of cladding on high rise offices and the shattering of glass in the windows of shops and public buildings. In analysing the results of this form of blast loading, it is usually assumed that the instantaneous pressures, and the pressure decay, are similar to those experienced by flat areas of masonry, brickwork or reinforced concrete.

8.2 CIVIL BRIDGES

Explosive damage to civil bridges has occurred in the past as a result of military operations, either in an attacking mode when for tactical reasons it has been necessary to demolish bridges by bombing, artillery shells, rockets or cruise missiles; or in a defensive situation where the demolition of a bridge prevented the enemy from advancing along a planned route. Defensive demolition was normally carried out by the emplacement of cutting charges on selected bridge members.

Bridge damage can also occur from non-military disasters such as vapour cloud explosions, tanker explosions following collisions, fuel explosions following the crashing of aircraft, or fuel explosions following earthquakes. Instances have been recorded of damage due to terrorist activities, mainly relatively slight damage to decks and truss members.

One of the most informative investigations into bomb damage (from aerial bombs) on large civil bridges was made by a UK Ministry of Home Security team in 1944 for the British Bombing Research Mission. When the Allied armies were fighting in Normandy after D-day heavy air attacks were made on road and rail bridges over the River Seine to prevent supplies reaching the German divisions. After the Allied advance over the Seine and on to the Rhine and its tributaries in September 1944, there was a chance to investigate the levels of damage to the Seine bridges. Many of the bridges (and the majority were rail bridges) were damaged three times during the Second World War; firstly when the French authorities demolished parts of bridges in 1940, secondly by the very severe bombing in mid-1944, and thirdly by the German army during the retreat from Normandy and Northern France in the autumn of 1944. The reports of the team of structural experts, originally secret, are now unclassified, and air photographs of the destruction have been released by the UK Public Record Office and reproduced by Her Majesty's Stationery Office. The report on the Seine bridges [8.19] was in four parts and one of the authors was Dr Frances Walley. Dr Walley, as a young scientist and engineer, carried out many research projects during the Second World War in support of the Research and Experiments Department of the UK Ministry of Home Security. Much of his outstanding work is still relevant, and he has recently published a valuable paper on the wartime efforts by himself and others in support of the need to understand the fundamentals of the effect of explosive forces on structures [8.20].

The Seine bridges report makes the general point that continuous girder bridges are very resistant to collapse by bombing, and that in many positions the main girders of such bridges can be severed without concomitant collapse of the entire structure. There is a usually generous allowance in design to cover redistribution of forces and moments due to cuts under dead load conditions only. To guarantee collapse it was usually necessary to cut two main booms in the same bay. It was also clear from the survey that the most vulnerable components of heavy girder bridges were the piers. Masonry arches were found to suffer greater local damage from single hits, but the spans were less and temporary repairs were more readily carried out. The number of hits was very low on the Seine bridges. For every 100 bombs released in all types of attack (high level and low level) three direct hits were scored, two on the superstructure, one on the supports.

The study concluded that for railway girder bridges the most efficient general purpose bombs were 2000 lb in total weight, and for road girder bridges 1000 lb bombs were more appropriate. For masonry arch bridges the effectiveness of the bombs varied with span, from 1000 lb bombs (0 to 50 ft), to 2000 lb (50 to 100 ft). Although there were instances of 500 lb bombs being effective between 0 and 50 ft span, the use of 1000 lb bombs was thought to be preferable. High-level attack with instantaneous fusing of the bombs was recommended for girder bridges, and low-level attack with delayed fusing for masonry arches.

The Bombing Research Mission ascertained the effect of bombs of known weight across the Seine between Paris and Rouen. Ten railway girder bridges, nine railway arches, five road girders and three road arches were examined. In terms of residual strength, rapid repair and speedy re-use, some of the bridges examined provide useful information. We will look at these in turn:

(a) d'Eauplet bridge (railway) at Rouen, Figure 8.2.
    The bombing history was as follows:

7 May 1944	70 × 1000 GP bombs	No damage
25 May 1994	54 × 500 GP bombs	No damage
25 May 1944	8 × 500 GP bombs	No damage
27 May 1944	45 × 1000 GP bombs	Pier 3 severely shaken
28 May 1944	46 × 1000 GP bombs	Pier 3 hit again, causing complete
	20 × 500 GP bombs	destruction.

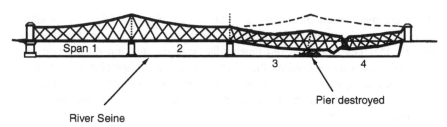

Figure 8.2 d'Eauplet railway bridge, 1944 (from Hill *et al.*, ref. 8.19).

The bridge was then used for road traffic by removing rails and sleepers from one track over spans 1, 2 and 3, and building a raised wooden roadway across the dip of damaged span 4. It is interesting that aerial reconnaissance by the Allied forces between May and August 1944 did not detect the destruction of the pier, and it was thought that all the bridge damage resulted from the cutting of the top booms. In fact, the top booms did not buckle until the pier collapsed.

(b) Tourville bridge (railway) 7 miles south of Rouen, Figure 8.3.
   The bombing history was as follows:

7 May 1944	21 × 500 bombs	Lower booms cut at 6 places in
	17 × 1000 GP bombs	spans 1 and 2
8 May 1944	48 × 1000 GP bombs	
9 May 1944	36 × 1000 GP bombs	Further cuts on top and lower
	36 × 1000 GP bombs	booms of spans 1 and 2
10 May 1944	64 × 1000 GP bombs	Span 1 collapsed completely, and 9 panels of span 2 fell into the river.

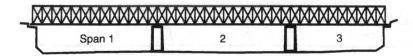

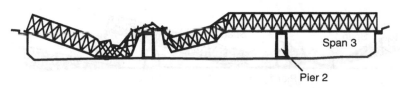

Figure 8.3 Tourville railway bridge, 1944 (from Hill *et al.*, ref. 8.19).

After the destruction of the bridge, Italian and Russian labourers under German control constructed a diversion bridge immediately on the downstream side. This was attacked repeatedly from the air, and during one of these raids a hit was scored on Pier 2 of the original bridge. Half of the pier collapsed and span 3 tilted downstream, interfering with the diversionary bridge. The German engineers therefore demolished the rest of the pier and dropped span 3 into the river.

(c) Le Manoir bridge (railway) 9.5 miles south of Rouen, Figure 8.4.
The bombing history was as follows:

21 May 1944 A single bomb fell on Pier 1, cutting the end post
               and overhead cross girder
25 May 1944 68 × 500 GP bombs Damage to main booms in spans 1
             36 × 500 GP bombs and 3
26 May 1944 32 × 500 GP bombs No further damage
27 May 1944 73 × 2000 GP bombs Span 1 hit near pier. End of pier destroyed allowing downstream girder to collapse. Left bank end of span 2 destroyed, and fell into river
30 May 1944 77 × 500 GP bombs No further damage
             16 × 1000 GP bombs
14 June 1944 Final air attack Span 3 cut and end of span near abutment fell into the river. The other span rotated about Pier 2, causing further damage.

The form of this destruction showed that it was possible to destroy a bridge of this type without completely demolishing a pier, as long as adjacent spans were completely severed.

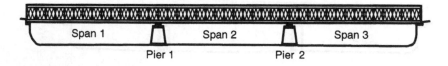

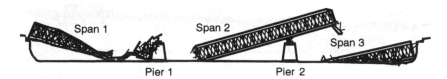

Figure 8.4 Le Manoir railway bridge, 1944 (from Hill *et al.*, ref. 8.19).

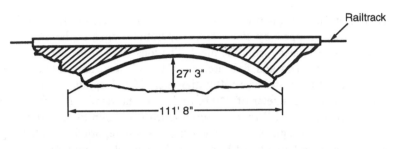

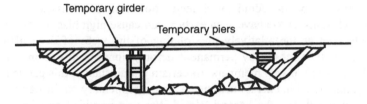

Figure 8.5 Vacoulers railway bridge, 1944 (from Hill *et al.*, ref. 8.19).

(d) Vacoulers bridge (railway) 30 miles north-west of Paris, Figure 8.5.
(A single span mass concrete arch bridge).
The bombing history was as follows:

3 June 1944	97 × 2000 GP bombs	No damage
11 June 1944		
12 June 1944	70 × 200 GP bombs	Bridge demolished, but temporary single line bridge erected in 18 days
20, 21, 24, 29 June	120 × 500 GP bombs	No damage during construction of temporary bridge
15, 19, 25, 31 July	143 × 1000 GP bombs	No damage to temporary bridge.
3, 8 August	774 × 500 GP bombs	

The ten attacks 20 June to 8 August during which 1037 bombs were dropped failed to damage the single line temporary bridge.

The explosive loading of the bridges described above was the result of direct hits of aerial bombs on structural members or piers, but it was realized towards the end of the Second World War that the damage resulting from near misses of very large bombs could often be more significant than damage due to direct hits from smaller weapons. Targets such as railway lines, viaducts and bridges were considered relatively invulnerable because of their extreme tenuity, but a potent way of destroying or disabling large lengths of bridges was to produce very large craters that destroyed the bridges foundations. As discussed earlier in 7.3, the UK Tallboy bomb, weighing 12 000 lb, was designed to attack German U-Boat pens on the Atlantic Coast of France, and this was followed by the

Tallboy (large) or Ground Slam bomb (22 000 lb). However, the first Tallboy (large) bomb) was dropped by the UK in operations against the Bielefeld Viaduct, and it fell 50 ft away from the piers. The loosening of the earth caused the destruction of six piers and five arches of the viaduct, and although the approximate shape of the crater was 150 ft diameter, 45 ft deep, the loosening of the soil occurred over a diameter of 240 ft, and to a total depth of 75 ft. These lessons of the Second World War were perhaps learned again by the forces of the west in the Gulf War against Iraq, when the US was reported to be turning away from the precision bombing of bridges with relatively small bombs to the use of very heavy bombs that could cause destruction from near misses.

Near miss explosions, as we have seen earlier, can cause high blast pressures on components such as the relatively thin plating of the webs of deep plate girders. Damage is then mostly by permanent deformation and distortion, and removal of the integrity of web members. In certain circumstances this can lead to shear failure and possible lateral buckling. A final commentary on the damage and repair of bridges during the Second World War was provided by Christopherson [8.3], who briefly examined ways of pre-strengthening existing civil bridges against the possibility of bomb damage. He emphasized the importance of permeability, ductility and redundancy, and the need to eliminate the possibility of progressive collapse in viaducts. In general, the results of surveys in the UK and US suggested that when it came to bombing from aircraft, it was better to shake down the piers than shoot up the superstructure. The susceptibility of bridge piers to undermining processes is, of course, well known from studies of civil bridge structural failure, where it has been found that about half of all bridge failures is due to scouring under piers and abutments.

The explosive damage to bridges from hand-placed contact charges is also an important aspect of our review. In 1971, for example, during a local war in Bangladesh, explosive charge damage was recorded on a number of relatively small reinforced concrete road bridges, and on one steel truss bridge. The damage was often in the form of holes in the deck, sometimes involving the partial destruction of multi-beams supporting the deck slabs, and occasionally this was severe enough to cause the complete collapse of a bridge span. Damage was also reported to reinforced concrete piers, which collapsed sideways, and to masonry abutments. The latter suffered extensive cracking and spalling and often a partial collapse of the bridge seating.

A famous example of bridge deck damage due to hand-placed explosives was the attack by German frogmen against the Nijmegen road bridge in September 1944. This has been reported in detail by Hamilton [8.21], who recounts that early in the morning of 28 September large explosions were heard. Over 70 feet of the roadway disappeared, following the actions of the frogmen, who used naval mines weighing 1200 lb, fitted with float chambers. These were placed in a necklace around the piers before the float chambers were released and delay fuses activated; however, the mines released their impact forces upwards, failing to damage the piers, but disrupting the roadway. It was necessary to build twin

Bailey bridges, each 80 ft in length, across the gap in the deck in order to continue to use the bridge for military traffic.

More recently there have been a number of demolitions, or part demolitions of civil bridges during the conflict in Bosnia. Reinforced concrete box section bridges have been destroyed or partially destroyed, mainly by charges placed at deck level; but there have been instances of damage to piers by explosives set at their bases. The problems encountered because of explosive damage to Bosnian bridges have been discussed by Pelton [8.22].

In order to classify the extent of explosive damage it has been suggested that the main load carrying components of all bridge structures can be classified as follows:

Class One: Complete destruction of the component will result in the com-
              plete destruction of the load-carrying capacity of the bridge. This
              could apply to suspension bridge towers and cables, cable-stayed
              bridge towers, single piers, single box girders or abutments.
Class Two: Complete or partial destruction of the component will seriously
              reduce the load-carrying capacity of the structure, but repair will
              allow a reduced capability to be maintained. This could apply to
              single main girders of bridges with at least two main girder
              systems or to the main arches of steel or concrete bridges with
              two or more parallel arches. It also applies to the deck slabs of
              composite bridges and to the cross girders of truss girder
              bridges.
Class Three: Complete or partial destruction of the component will only have
              a localized effect on load-carrying capacity, and repair will allow
              the full capability to be restored. This could apply to the deck
              slabs of non-composite bridges or to single members in highly
              redundant trusses or frameworks.

The calculation of the residual strength of bridges damaged by explosions or by penetration and fragmentation of explosive-carrying missiles can be based on safety levels and on the specification of live load capacity as an equivalent uniformly distributed load. The chances of crossing a damaged bridge safely before any repairs are made can be related to five levels of risk, each level being linked to the number of vehicles of maximum load class that can cross the damaged structure before it becomes completely unsafe. The decrease in safety with time can be linked to the growth of deep cracks around the edge of a damaged area in a concrete slab deck or to the growth of tearing in a steel plate girder that has been pierced by fragments; or to the gradual onset of instability in a supporting pier due to settlement into a nearby crater.

If the bridge fails completely after 10^n crossings of the heaviest vehicles, the safety level can be expressed in terms of the value of n, as follows:

198   *The effects of explosive loading*

Safety Level 0: $n = 0$ Extreme risk. Probability of failure as the bridge is crossed by one vehicle of maximum load class.

Level 1: $n = 1$ High risk. Probability of bridge failure after 10 crossings of maximum load class.

Level 2: $n = 2$ Moderate risk, with bridge failure probable after 100 crossings.

Level 3: $n = 3$ Low risk, with failure after 1000 crossings.

Level 4: $n = 4$ Minimal risk, with failure after 10 000 crossings.

(The minimal risk level might well be of the same order as the annual probability of failure of an elderly undamaged civil bridge under maximum road class traffic.)

The design loadings for civil bridges were compared in 1979 by the Organization of Economic Co-operation and Development (Paris) [8.23] who conducted a study in which the total bending moments caused by live loading in various codes of practice were calculated for a simply supported bridge. Calculations were made for two, three and four traffic lanes over spans from 10 m to 100 m, taking account of impact and reductions for multiple lane loading. Differences in allowable stress levels were taken into account. The maximum bending moments obtained were converted into an equivalent uniformly distributed load q_L. Consider, as an example, the results for American AASHTO load specifications applied to road bridges on motorways, trunk roads and principal highways (A roads), given in Table 8.4.

From these figures it is possible to calculate the maximum live load bending moment that can be applied to an undamaged bridge under working conditions ($=q_L L \frac{2}{8}$ for a simply supported span of length L), or under ultimate conditions ($=1.5 q_L L \frac{2}{8}$). The corresponding maximum shear forces would be $q_L L/2$ and $1.5 q_L L/2$.

When assessing residual strength, a knowledge of the live loading, q_L, is not enough. It is necessary to know the equivalent dead load per metre ($=q_D$) due to the self weight of the bridge and its accessories. Various equations have been proposed for the rough calculation of total dead load, and these depend on the number of traffic lanes, an estimate of the load classification and properties of

Table 8.4 Values of equivalent uniformly distributed load, q_L, for AASHTO loading

Two lane bridges:	Span in metres	20	40	60	80	100
	q_L in KN/m	110	62	50	46	42
Three lane bridges:	q_L in KN/mm	150	82	66	63	60
Four lane bridges:	q_L in KN/m	170	92	73	67	64

The values of q_L to cause the bridges to collapse could be taken approximately as 1.5 times those given in the table.

the material from which the bridge is constructed. An approximate relationship, used by defence engineers for spans between 10 and 100 m, is

$$q_D = K(L + 50)(1 + 2m)(30 + c)/800 \, KN/m, \qquad (8.2)$$

where m is the number of traffic lanes, c is an estimate of the military load class, and K depends on the materials from which the bridge is constructed. For a steel main structure and timber deck, $K = 1$; for a steel main structure and concrete deck, $K = 1.25$; for prestressed concrete bridge $K = 2.4$ and for other reinforced concrete bridges $K = 4.0$.

As an example, the dead load per metre for a reinforced concrete bridge having two lanes of traffic and a military load class of 70 would be

$$q_D = 4(L + 50)5(30 + 70)/800 \, KN/m, \qquad (8.3)$$

so that for a simply supported span of 50 metres, $q_D = 50 \, KN/n$. This is of a similar order to q_L from Table 8.4. For a four lane bridge having a similar span q_D would rise to 90 KN/m, again of a similar order to q_L.

According to Eq. (8.2) the dead load per metre for a reinforced concrete bridge is 2.2 times that for a steel and concrete bridge having a similar capacity, width and span, and 1.7 times that for a prestressed concrete bridge. These figures are independent of span within the range given and imply that the effect of prestressing a bridge is to reduce its dead weight for the same span and loading by 40%. This is consistent with the notion that prestressing is broadly equivalent to applying a distributed upward vertical force on the girder.

If the girder weight is reduced by 40% of its non-prestressed value, then for a rectangular girder section of constant width the effect will be a reduction in section depth. This in turn leads to a reduced capacity to carry a bending moment to about 36% of its original value. If the section is an I beam or hollow box, the reduction would be to about 60% of the original value. For a practical bridge girder section the true reduction in moment capacity would probably be around 50%.

This infers that if the prestressing tendons are cut by an explosion the total capacity of the girder to carry live plus dead loading could be halved. We have already seen that the dead load moment for concrete bridges of less than 40 metres span could well be equal to the dead load moment under working conditions, so after the destruction of tendons the girder would just take its dead load with a margin of safety of about 1.5. An application of live load equal to about one half maximum would then cause complete failure, and in fact any application of live load would begin to cause undesirable permanent bending deflections and structural deterioration.

As the length of a single span girder bridge increases, the proportion of the moment capacity used to support the dead load will also increase. In the limit, for very long suspension or cable stayed bridges up to a single span of 1500 m, it is often assumed that 80% of the total loading is due to the deadweight of the deck structure and stiffening girders. This figure has also been shown to apply to

Table 8.5

Span in metres	0	20	40	60	80	100
Dead load/total load bending moments	0	0.48	0.70	0.80	0.86	0.87

prestressed concrete bridges in the 100 m to 200 m span range, and it has been suggested that the ratio of dead load to total (dead plus live) bending moments varies with span according to Table 8.5.

If we take the above results to apply in general to civil bridges, they show that a relatively small amount of explosive damage to a key part of the structure of a long span bridge will reduce the strength of the bridge such that it is only just able to cope with its own dead load. In this event, the chances of being able to re-classify the damaged structure to take even a very light vehicle are not high.

The residual strength of any bridge component, whether class one, two or three, is often linked to the amount of cross-sectional area that is missing, where the reduced area is located, and how far the reduced area extends along the member. It can also be linked to the degree of deformation suffered from blast due to near misses, which may be in the form of permanent lateral deflection, and shortening or twist. The residual strength of groups of components can also be influenced by the complete destruction of a single component or of a joint between adjacent components.

When the explosive damage results in the loss of cross section, the weakening may result in a reduction in tensile, compressive, bending or shear strength, or to combinations of these. The loss of section can, of course, take any shape, but let us first assume that the width, b, of a rectangular section is maintained, but that the damage results in a reduction of the depth, d, as shown in Figure 8.6.

The type of component can be linked with the order of damage by the relationship:

$$\frac{\text{Residual strength}}{\text{Initial strength}} \propto \left[\frac{\text{Residual area}}{\text{Initial area}}\right]^{m}, \tag{8.4}$$

where the index m can be 1, 2 or 3. The first term in Eq. (8.4) is sometimes referred to as the 'strength ratio', and the second as the (area ratio)^{m}.

For a simple tension member it can be assumed that the local loss of cross section extends over the whole length of the member, and then $m = 1$, because tensile strength is directly proportional to section area (neglecting stress concentration, crack growth or fatigue). For a member such as a girder web, or the vertical plating of a box girder, when damage results in local section loss near an area of maximum shear, it is approximately correct to take $m = 1$.

For members contributing to bending strength, it can again be assumed that the damage extends over the whole of the beam length. In this event $m = 2$, because the section modulus will generally be proportional to the square of the depth. For a simple slender compression member such as the top boom of a

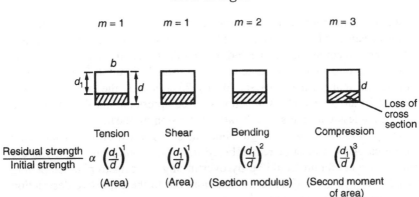

Figure 8.6 Relationship between residual strength and initial strength for a given loss of depth of a rectangular section.

truss, or the pylon of a cable-stayed bridge, assume that the section loss extends over the entire length. This leads to $m = 3$, because the second moment of area, required in the calculation of buckling or compressive strength, will be approximately proportional to the cube of the depth of the component.

For a very stocky compression member such as a masonry pier or abutment, damage will only lead to $m = 3$ if so much of the cross section is removed that the remainder buckles as a slender strut. Otherwise, the order is more likely to be $m = 1$. The above discussion is illustrated in Figure 8.6, and from it we can form a relationship between strength ratio and area ratio, as shown in Figure 8.7. This figure represents the simplest loading and the most elementary member

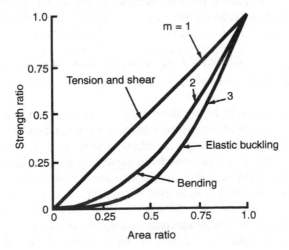

Figure 8.7 Relationship between strength ratio and area ratio. Permissible stress design in the elastic range.

shape, and any variations such as combined loading or unusual cross sections will result in values of *m* between 1 and 3 in the strength ratio equation. For example, a bridge component subjected to combined bending and compression might be expected to have an *m* value between 2 and 3. Figure 8.7 indicates that for damage that reduces the cross section by 25% or less, the strength ratio is very similar for simple members in bending or compression, but this ratio could in turn be 25% less than the strength ratio in tension or shear.

This analysis looks to have a logic about it, but is based on rather idealistic circumstances. Also, it must be remembered that if the foundations of a pier are undermined, bridge failure could be by overturning – a catastrophic happening not linked to bridge ductility or loss of section, but to the explosive destruction of external support from the ground.

The other extreme to the damage configuration assumed in Figure 8.6 would be the vertical slicing of the section resulting in the reduction of the width from b to b_1. In all the loading cases considered the value of *m* would now be 1, and the strength ratio would be directly proportional to the area ratio. In practice cross-sectional damage would probably be somewhere between the two extremes discussed above.

The assessment of strength reduction from a knowledge of the area reduction of truss members was pursued during and after the Second World War by the designers of the military 'Bailey Bridge' in the UK. Their results were presented in the form of a table, given in reference [8.24], which gave a relationship between the amount of damage to the webs and flanges of *I* beams and percentage residual strength of verticals and diagonals in the bridge truss panels. For example, the complete removal of the web in shorter members led to a residual strength of 66%, and removal of half the web gave 73%. Complete removal of one flange gave a residual strength of 24%, and removal of one flange and half the other reduced the figure to 18%. Similar residual strengths were given for main chords consisting of two back-to-back channels. The removal of the web of the channel gave a residual percentage of 63% for the shorter members, and removal of the webs of both channels gave 61%. Complete removal of one channel flange and web gave a residual percentage of only 3%.

The article in reference [8.24] on the damage assessment of military bridges noted that in the majority of cases where bridges were damaged by enemy air attack or shell fire, several bridge members would have been reduced in strength. Each member was then considered separately and the final classification assessed on the worst case. The article also stated that if a bridge member is struck by flying metal and is deformed but not holed, it must be 'carefully watched as loads cross the bridge'. If further deformation occurs, the member is treated as severed, and the bridge strength assessed accordingly.

Rapid methods of evaluating residual strength can be illustrated by taking examples, and three areas where damage might occur to Class One bridge components are:

(a) damage to deck slabs that carry longitudinal bending moments;
(b) damage to main longitudinal side or multi-deck girders;
(c) damage to main longitudinal box girders.

These will be considered in turn:

(a) Deck slabs carrying longitudinal bending loads

These are common in short span reinforced concrete slabs that carry loads across gaps in the 0 to 40 m range. Some have a constant transverse section, others have variable sections, haunches, and prestressed or post-tensioned reinforcement.

Suppose an explosion has caused a jagged hole and severed reinforcement, as shown in Figure 8.8, and that a rapid assessment of trafficability prior to repair is required. The way of going about this is to mark the transverse limits of the hole from the edge of the roadway (not including footwalks), and extend these limits longitudinally to form the lines CD and EF in the figure. Any traffic crossing the damaged bridge should cross within the zone ABCD, and a check is needed to ensure that there is a clear undamaged load transfer path from the deck to the supporting structure. Beyond the immediate visible boundaries of the damaged areas there could be internal cracks in the slab. These would spread under the influence of cyclic loads and might eventually reach a distance of up to two metres on either side of the damage boundary.

If only one crossing is going to be made before the bridge is closed (i.e. 10° crossings), the width available for traffic is equal to $\beta W - \delta W/2$ metres, as indicated in Figure 8.9. This could be thought of as a safety level of 0, and the available section to resist bending at the position of the hole would have a total width of $W - \delta W$ metres. If we consider that the internal cracked zone will spread uniformly, the relationship between safety level, numbers of crossings of the maximum load class vehicles and the total cracked width is as shown in

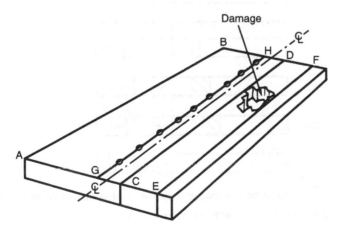

Figure 8.8 Damage to the deck slab in a short-span reinforced concrete bridge.

Figure 8.9. If it is intended, for example, that 100 full load class vehicles should be able to cross the damaged structure before repairs are commenced, residual strength calculations should be based on a width at the hole of $W - (\delta W + 2)$ metres, and a maximum width of load of $\beta W - \delta W/2 - 1.0$ metres.

If the damaged area lies close to the ends of the span, ie between 0 and $L/8$ or $7/8L$ and L, the remaining shear capacity should be checked. The working shear strength of the undamaged bridge is $q_L L/2$, and this would be supplied by a section of full width W. If the reduction in shear area at the hole for 100 crossings is proportioned to $[W - (\delta W + 2)]/W$, the shear strength at the ends of a damaged bridge would be this fraction multiplied by $q_L L/2$.

(b) Main longitudinal girders (multi-girders or pairs of side girders)

If reinforced concrete beams in a multi-beam construction associated with a deck slab suffer any damage, it is best to assume that the particular beam does not contribute to the overall bending or shear resistance of the bridge.

If steel beams suffer flange damage, it is prudent to assume that they make no contribution to bending resistance, but if only minor web damage recurs (from air attack by cannon shells, for example) then deterioration of web shear strength will not be crucial as long as the holes and distortions do not occupy more than 20% of the total side area between transverse stiffeners. In the

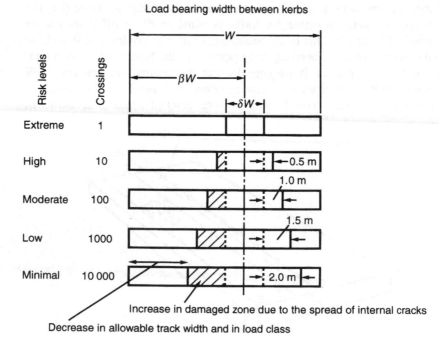

Figure 8.9 Allowable track width for a desired level of risk.

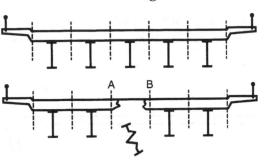

Figure 8.10 Damage to the central girder and associated deck in a multi-beam bridge.

absence of stiffeners the side area considered should be that enclosed by a length of beam equal to its depth.

If a load-carrying roadway between kerbs is supported by n equally loaded longitudinal beams, then a rapid assessment of the reduced bending strength ratio will be $(n - 1)/n$. To take a dimensional example, suppose a 20 m span bridge with two lane capacity (width 10 m) on a trunk road, consists of a concrete deck slab supported by 5 equal steel beams, and that one of the beams is seriously damaged as the result of an explosion on the deck immediately above it, by the passage of a missile, or by the effects of a stand-off explosion underneath. Suppose that the centre beam has suffered the damage, as indicated in Figure 8.10. Taking the section of deck AB out of the strength calculations, as well as the I beam, will reduce the capacity by about 20%, and also leave a reduced deck space for the passage of vehicular traffic. It is necessary then to check that the reduced bending strength and the reduced passage width are compatible. This assumes that there is no redistribution of loads between beams.

Thus, for a span of 20 metres, $q_L = 110\ KN/m$, and if it assumed that for this type of construction and span $q_D/(q_L + q_D) = 0.5$, then $q_D = q_L = 110\ KN/m$, and $q_D + q_L = 220\ KN/m$. The reduction because of the damage given in Figure 8.10 leads to a capacity of $0.8 \times 220 = 176\ KN/m$. Subtracting the virtually unchanged dead loading leaves $q_L = 66\ KN/m$. The deck width available for loads that give this live loading is about 4 metres, and this is more than required since the maximum width of load to produce $q_L = 66\ KN/m$ is 3.4 metres.

(c) Main longitudinal box girders

If the structure between simple bridge supports is a single or multi-cell box girder, it is probable that the bridge span is over 60 m, and could be as much as 200 m. At these spans the values of q_L do not change much with increases in length. For example, a bridge 100 m long, designed to carry four lanes of motorway, could have a total load-bearing width of 20 m, as shown in Figure 8.11 The value given for q_L is $64\ KN/m$ for the AASHTO loading of a civil bridge, and at this span the value of q_D is at least 80% of the total, so that

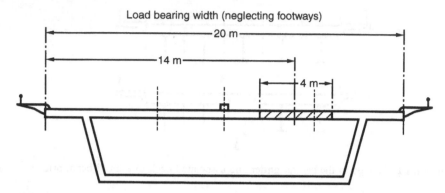

Figure 8.11 Example of damage to the deck slab of a single cell box girder bridge.

$q_D = 256 \, KN/m$. The live load moment = 80 000 KN/m, the dead load moment = 320 000 KN/m, and the total 400 000 KN/m. Suppose an explosion occurs on the deck of a single call box girder concrete bridge, leaving a hole 4 m wide, the centre of which is 14 m from the edge of the footway; then it can be shown that the residual bending strength for 100 subsequent crossings (level 2 risk) is about 340 000 KN/m. Since the dead load moment is virtually unchanged, the remaining live load capacity is only 20 000 KN/m. This is only 25% of the original undamaged capacity.

An examination of explosive damage to bridges by the author has indicated that a method of rapid field assessment of residual strengths can be derived and presented as a graphic display on a personal computer. This would show the damaged area, zone of internal cracking, available residual lane width, the load class of vehicle that would still use the bridge, crossing speed limitation and number of crossings that could be safely made.

There are at least 370 different types of bridge structure in the world, and for each of these there are at least twelve sources of explosion and twelve records relating the point of the explosion to the structure. The sources influence the type of damage and the explosion points influence the extent of the damage. This gives a total of about 53 000 separate groups of calculations, many of which would be very similar.

To take one example, Figure 8.12 shows the possible explosive sources for damage to a tower and cable system, such as part of the structure of a cable-stayed bridge. The extent of the damage from each source can be used to calculate the safe residual strengths of the main load-carrying components. We must limit the discussion to these general remarks, because more detailed information could be of a restricted nature.

It is acknowledged that the major part of section 8.2 has been drawn from a report by the author to the Defence Research Agency (Chertsey), who commissioned a study on the assessment of damaged civil bridges.

8.3 AIRCRAFT AND SHIPS

Damage to military aircraft from explosive blast or penetrating projectiles has obviously been a subject of structural interest since the First World War, but military security prevented much discussion in the open literature. One of the most valuable sources of unclassified reporting of research and analysis in this field did not appear until after the Second World War, under the auspices of the North Atlantic Treaty Organization. The Advisory Group for Aerospace Research and Development (AGARD) was established so that panels of experts could provide scientific and technical assistance to member nations, and one of these, the Structures and Materials panel, was much concerned in the 1970s with the physical vulnerability of aircraft. Among the subjects covered were damage from projectiles and blast, the failure characteristics of the structure, residual

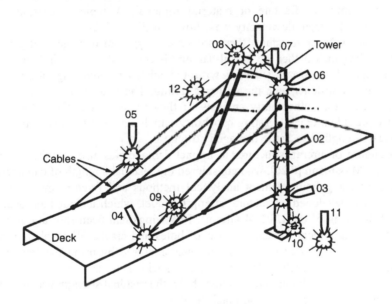

01 Direct hit on portal frame member
02 Direct hit on tower leg from horizontally propelled missile
03 Direct hit on pier leg from horizontally propelled missile
04 Direct hit on connector between cable and deck
05 Direct hit on cable
06 Direct hit on connector between cable and tower
07 Partial penetration of warhead with no explosion
08 Contact explosion on portal frame from static or shaped charge
09 Contact explosion on cable from static or shaped charge
10 Contact explosion on pier leg from static or shaped charge
11 Near miss explosion at ground or water level
12 Stand-off explosion near main portal frame

Figure 8.12 Possible sources of damage to a cable stayed bridge.

strength and life after damage – all part of the study of the impact damage tolerance of aircraft structures.

It was realized that existing design guidelines and specifications did not fully address the projectile damage threat, and to improve design methods Avery, Porter and Lauzze reviewed structural integrity requirements [8.25] and resistance to battle damage. The damage in metal sheet, stiffener and plate structures typical of modern aircraft takes the form of cracks, spallation, petals, holes, dents or gouges, and the authors pointed out that for a given target material the type of damage depends on sheet thickness, projectile velocity and impact angle. A 'damage regime' diagram for 0.30 armour piercing projectiles in plating made from the aluminium alloy 7075-T6, taken from ref. [8.25] was shown in Figure 7.23. The most useful way of quantifying projectile damage is by the 'lateral damage' notion, which is defined as the diameter of an imaginary circle that just encloses the limits of fracture or material removal. A typical variation of damage size with projectile velocity was shown in Figure 7.24.

In skin and stiffener structures there are several types of damage configurations, depending on whether the lateral damage is confined to the skin between stiffeners, spreads across a stiffener with the stiffener remaining intact, or spreads across a stiffener with the stiffener failing. The latter type is normally the critical case for vulnerability analysis. If they remain intact the stiffeners frequently provide a crack-arresting capability which can significantly improve the residual strength of a battle-damaged structure.

The response of aircraft structures to impact damage was also examined, for example, by Massmann [8.26], who considered the residual strength of damaged structures by using finite element analytical methods and fracture mechanics techniques. He developed a structural strength model, which included an idealization of the wing of the US F84 aircraft. F84 wings had been shot at during field tests, giving typical damage patterns of holes and cracks. Wing plating in which the damage resulted in cracks only was analysed by using the fracture toughness theory of Griffith, Westerg and associates to determine residual strength. When the crack ran into a circular hole, the residual strength was found to be a function of hole radius and crack length.

Test results and analysis showed that hits close to the front spar of the wing considerably reduced the load capacity, and that wings constructed of stiffened panels and three spars had a residual strength of over 2.5 times the residual strength of milled integral panels having two spars (both geometries having been designed to have similar initial load capacities). The stiffened panel design was greater in weight than the integrally milled design, but in terms of the entire weight of the aircraft, the increase in weight was less than 1% and the decrease in manoeuvrability negligible.

In addition to the possible loss of strength from impact damage, the stiffness of the aircraft structure may be altered, and stiffness degradation can lead to aero-elastic problems associated with flutter, control and load redistribution. Aircraft structural design to increase resistance to blast or penetrative damage was reviewed in 1972 by Avery, Porter and Walter of the Boeing Company of Seattle [8.27]. They pointed out that if the damaged structure survives intact after the explosion, it will be subjected to further loading as the flight continues. Residual strength is therefore a function of time, since cyclic loading will continue to extend the damage because of fatigue. In their paper, Avery *et al.* referred to Figure 8.13, which compared critical gross stress for stiffener and skin failures and for a typical structural configuration.

In recent years there have been instances of aircraft surviving on-board explosions and subsequently landing with holes through the fuselage and decompression of the main cabin. There have also been instances of a single explosion so severely damaging the structure of an airliner that it disintegrated rapidly and crashed with catastrophic lose of life. One of the most harrowing disasters occurred at Lockerbie in Scotland, in December 1988 when a Boeing 747 of the Pan Am company suffered an internal bomb explosion in the forward

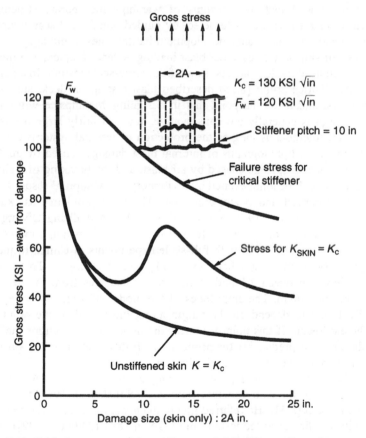

Figure 8.13 Critical gross stress for stiffener and skin failures (from Avery *et al.*, ref. 8.27).

hold, and fell from the sky. The loss of this aircraft has led to research to investigate the blast forces on aircraft stiffened sheeting and structural members after the blast has travelled through various irregular geometries of spaces, conduits and ducts on its way from the hold to the vulnerable zones of the airframe or control equipment. The geometry of the problem lends itself to a finite element solution, but the basic variations of blast pressure with distance and configurations still employ the fundamental science discussed in the later sections of Chapter 5.

Turning now to ship structures, it is well known that below the waterline marine vessels are particularly vulnerable to torpedoes and mines, and above the waterline to shells, bombs and guided missiles. Much research has been devoted to these problems, and in addition there was considerable activity in the 1950s and 1960s to examine the response of naval vessels to nuclear blast. The work of Hopkinson and others on anti-torpedo blister structures during the First World War has already been mentioned, and at the more recent end of the research spectrum it is noticed that new designs of warships incorporate structural features that make them less easy to detect by guided bombs and search/strike missiles. The costly attack/defence development battle is never ending.

The analysis of ship structures under blast loading is very complex, and much of the knowledge used by designers is drawn from large-scale testing. In a paper given in 1988, Charles Smith [8.28] described blast testing on deck, side and bow components, and made the point that in designing warships to withstand blast elastic design is generally too conservative, and that fairly large inelastic deformations are tolerable as long as the protection of internal systems can be maintained. He showed photographs of internal blast damage caused by guided missiles, superstructure damage caused by air blast, and the breaking of a ships back amidships by hull bending excited by a vibratory or 'whipping' response of the hull to a non-contact underwater explosion. The latter problem was examined in a paper by Hicks in 1986 [8.29], and by Jinhua and Zhang Qiyong in 1984 [8.30]. The latter authors derived formulae for dynamic bending moments which were shown to agree well with full-scale experiments in China. In these tests charges varying in weight between 200 kg and 1000 kg of TNT were detonated at depths between 7.5 and 40 m, and at distances from the vessels between 7.5 m and 100 m. The amplitudes of the modes of vibration were not unexpectedly found to depend on the length along the vessel of the point of action of the explosion. If this point was near the middle cross sections the 1st mode of vibration was predictably the prime mode. If the point of action was too near the quarter point the 2nd mode of vibration dominated.

Probably the most interesting paper referring to the history of underwater explosion research in the ship structure field was published in 1961, and was written by A. H. Kiel [8.31]. He referred first to the history of systematic testing, beginning with the first reported tests in the USA in 1881 (Abbott, [8.32]), and to the development between the two world wars of the side protection systems against torpedoes used in the battle ships *Midway* (USA), *Hood* (UK), *Tirpitz* and *Bismark* (Germany). He then reviewed the rapid development after the Second World War of our understanding of the effects of explosions on ship structures, noting particularly the work of King [8.33] and Hollyer [8.34] reported to the US Society of Naval Architects and Marine Engineers. Hollyer treated the important subject of the deformation of plane plates under shock loading.

Keil discussed damage patterns resulting from a torpedo or mine exploding in contact with the hull of a ship, an action that usually results in the tearing of a hole about 10 to 15 metres length in the fore and aft direction. If this causes rupture of the bulkheads adjacent to the hole, flooding will spread rapidly. However, the structural damage does not normally extend far, either in board or fore and aft. A typical rupture pattern is shown in Figure 8.14, taken from Keil's paper. If the explosion occurs at a stand-off from the side of the ship the shell might not rupture, but the plating of the hull could suffer severe plastic deformation.

In the discussion of Keil's work, an interesting contribution was made by Garzke and Dulin, who pointed out that considerable research was carried out between the two world wars by the Japanese Navy. In 1924 underwater contact explosion tests were carried out on the hull of an incomplete battleship, the

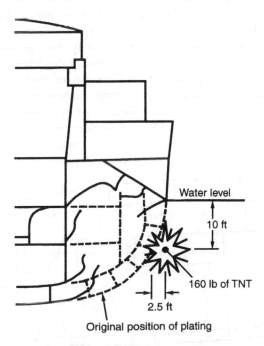

Figure 8.14 Internal damage to a hull from a simulated torpedo explosion (from Keil, 8.31).

Tosa, using torpedoes. The first hit amidships was with a weapon containing 350 kg of TNT, detonating 16 feet below the water line. This ripped a hole 20 ft by 18 ft and flooded damaged spaces with 1140 tons of seawater. The second test was to detonate 100 kg of TNT near the relatively unprotected bow and 12 feet below the waterline, and this ripped a hole 19.5 ft by 16.25 ft in the hull. A third test used 150 kg of TNT 20.8 ft below the waterline in the area near the main gun magazines, producing a hole 19 ft by 17 ft. It is interesting to note that the large variation in charge size did not produce a noticeable variation in the size of the hole through the hull.

A further useful test was carried out on the *Tosa*, when a projectile with an underwater trajectory was fired at the hull. The projectile was a 40 cm armour-piercing shell, which struck the ship 11.7 ft below the waterline, as shown in Figure 8.15. It penetrated the outer hull plate, the inner bottom plate and a torpedo bulkhead (3 inch high tensile steel plate) before exploding. A section 14 ft by 8 ft was carved out of the double hull, and a total of 2950 tons of seawater gained access to the damaged area. These tests have been recorded in the *History of Japanese Naval Construction*, written in the 1950s by 50 of Japan's leading naval architects and naval constructors.

The structural response to explosions of submarines is another area where much classified research has taken place. The resistance of hulls has been increased, partly by the introduction of high-strength steels which lessened the chances of hull splitting and flooding under attack by depth charges. It was pointed out by Roseborough, also in the discussion of Kiel's paper, that damage to internal components and systems was more likely than hull damage. Such damage could make vital controls inoperative, causing the submarine to exceed the hull collapse depth or conversely to surface out of control. The need to preserve the integrity of seawater systems within submarines, and their hull valves, in the face of explosions was also underlined by Roseborough.

A submarine hull is often a ring stiffened cylinder, which lends itself to an analytical as well as an experimental evaluation of structural performance, so

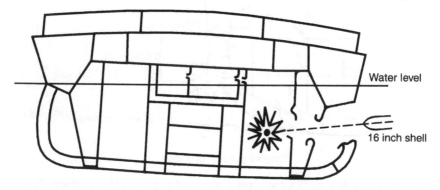

Figure 8.15 *Tosa* armour-piercing shell experiment (from Keil, ref. 8.31).

much time has been devoted over the years to the prediction of elastic response to transversely exponentially decaying shockwaves. Work in this field was summarized by Haxton and Haywood [8.35] in 1986.

8.4 REFERENCES

8.1 Thomas, W. N. (1948) The effects of impulsive forces on materials and structural members, *The Civil Engineer in War*, UK Institution of Civil Engineers, London, p. 65.

8.2 Baker, J. F., Leader Williams, E. and Lax, P. (1948) The design of framed buildings against high explosive bombs, *The Civil Engineer in War*, UK Institution of Civil Engineers, London, p. 80.

8.3 Christopherson, D. G. (1946) *Structural Defence*, UK Ministry of Home Security, Civil Defence Research Committee paper RC 450.

8.4 Eytan, R. (1992) Response of real structures to blast loadings – the Israeli experience. In P. S. Bulson (ed.), *Structures under Shock and Impact* (2), CM Publications and Thomas Telford.

8.5 Rhodes, P. S. (1974) The structural assessment of buildings subjected to bomb damage, *The Structural Engineer*, **52**.

8.6 Williams, M. S., Lok, T. S. and Mays, G. C. (1989) Structural assessment of damaged buildings. In P. S. Bulson (ed.), *Structures under Shock and Impact* (1), CM Publications and Elsevier.

8.7 UK Ministry of Home Security (1941) *Tests on Brick Surface Shelters at Stewartby*, Report RC 217, UK Ministry of Home Security, R and E Dept, April.

8.8 UK Ministry of Home Security (1941) *Report on Tests: Methods of Strengthening Brick Surface Shelters*, Report RC 272, Ministry of Home Security, R and E Dept, November.

8.9 UK Ministry of Home Security (1942) *Report on Tests on Four Factory-type Brick Surface Shelters*, Richmond Park, Ministry of Home Security, Group 6, Report RC 355, R and E Dept, October.

8.10 Walley, F. (1942) *Movement of Shelters due to Earthshock*, UK Ministry of Home Security, R and E Dept, Report RC 287, January.

8.11 US Army (1965) *Fundamentals of Protective Design (non-nuclear)*, US Army TM5-855-1, November.

8.12 Bernai, J. D. (1942) *Bulletin A-4*, UK Ministry of Home Security, R & E Dept.

8.13 Rempel, J. R. and Beck, J. E. (1985) Stability of walls against airblast forces, *Proc. 2nd Symp. on the Interaction of Non-nuclear Munitions with Structures*, Panama City Beach, Florida, USA, April.

8.14 Napadensky, H. and Longinow, A. (1985) Explosion damage to urban structures at low overpressure, *Proc. 2nd Symp. on the Interaction of Non-nuclear Munitions with Structures*, Panama City Beach, Florida, USA, April.

8.15 Ellis, B. R. and Crowhurst, D. (1989) On the elastic response of panels to gas explosions. In P. S. Bulson (ed.), *Structures under Shock and Impact* (1), CM Publications and Elsevier.

8.16 Ellis, B. R. and Crowhurst, D. (1991) The response of several LPS maisonettes to small gas explosions, *Conf. on Structural Design for Hazardous Loads*, Brighton, England, April.

8.17 Moore, J. F. A. (1983) *The Incidence of Accidental Loadings in Buildings 1971–1981*, UK Building Research Establishment, Current Paper 2/83.

8.18 *The Effect of Impact and Explosions* (1946) NRDC Summary Technical Report, Washington.

8.19 Hill, L. G. W. *et al.* (1944) *Report on Material Damage Caused by Bombs to Bridges*, UK Ministry of Home Security Research and Experiments Dept., Report RC 448.

8.20 Walley, F. (1994) The effect of explosions on structures, *Proc. Inst. Civ. Engrs Structs and Bldgs*, **104**, August, p. 325.

8.21 Hamilton, H. G. W. (1994) Reminiscences of a Corps SORE II in the British Liberation Army, 1944, *The Royal Engineers Journal*, **108**(3), December, 289.

8.22 Pelton, J. F. (1993) Bridge inspections in Bosnia – Operation Grapple, *The Royal Engineers Journal*, **10**(2), August.

8.23 Organisation for Economic Co-operation and Development (1979) *Evaluation of Load Carrying Capacity of Bridges*, Road Research Group Report, December.

8.24 *Military Engineering*, Vol III, published by UK Royal Engineers.

8.25 Avery, S. G., Porter, T. R. and Lauzze, R. W. (1976) Structural integrity requirements for projectile impact damage – an overview, *Proc. Conf. on Impact Damage Tolerance of Structures*, AGARD Conf. Proc. No. 186.

8.26 Massmann, S. (1975) *Structural Response to Impact Damage*, AGARD Report 633.

8.27 Avery, S. G., Porter, T. R. and Walter, R. W. (1972) *Designing Aircraft Structures for Resistance and Tolerance to Battle Damage*, AIAA 4th Aircraft Design, Flight Test and Operations Meeting, Los Angeles, California, August (AIAA paper 72-773).

8.28 Smith, C. S. (1988) The testing of marine structures, Proc. of Conf.: *The Future of Structural Testing*, Bristol University, Computational Mechanics and McGraw Hill.

8.29 Hicks, A. N. (1986) Explosion induced hull whipping, Proc. of Conf.: *Advances in Marine Structures*, ARE Dunfermline, May, Elsevier Applied Science Publications.

8.30 Jinhua, M. A. and Qiyong, Zhang (1984) The estimation of dynamic bending moment for a ship subjected to underwater, non-contact explosions, *Proc. Int. Symp. Mine Warfare Vessels and Systems*, London, June.

8.31 Keil, A. H. (1961) *The Response of Ships to Underwater Explosions*, US Society of Naval Architects and Marine Engineers, Annual Meeting, November.

8.32 Abbott, H. L. (1881) *Report upon Experiments and Investigations to Develop a System of Submarine Mines for Defending the Harbours*, Papers of the Corps of Engineers, No. 23.

8.33 King, R. W. (1959) *Modern Weapons and Ship Protection*, Paper to Chesapeake Section, SNAME, February.

8.34 Hollyer, R. S. (1959) *Direct Shock Wave Damage to Merchant Ships from Non-contact Underwater Explosives*, Paper to Hampton Roads Section, SNAME, April.

8.35 Haxton, R. S. and Haywood, J. H. (1986) Linear elastic response of a ring-stiffened cylinder to underwater explosion loading, Proc. of Conf.: *Advances in Marine Structures*, ARE Dunfermline, Scotland, May.

9

Response, safety and evolution

9.1 INTRODUCTION

Although the text of this book deals mainly with explosive loads and their effects, it would be an omission if we did not consider in a general way the analytical methods that seek to predict how structures respond to blast. We should also consider briefly the qualities of structures that are designed to resist known levels of explosive loading, because cases exist where the behaviour of the structure under shock loading can change the applied transient loading during the loading process.

Most hardened land structures are designed for military purposes, although some thought has to be given to response in the design of grain silos, chemical and explosive production plants, power stations and offshore structure blast walls. Any historical review has an early start, because permanent defensive installations have been an operational requirement since the beginning of civilization, and we are all familiar with the great defensive ditches, earth banks and stone walls built throughout the world in earlier ages. It is interesting to note that the Brits, the Celts and the Saxons tended to dig their fortifications, whereas the Romans and Chinese built walls.

As the knowledge of materials and structures grew, so the fortifications became more sophisticated, and stone-built castles became an early example of a hardened structure capable of protecting the inhabitants from attack for long periods, and then becoming the springboard for offensive operations. The design of castles, like all hardened structures, was conditioned by the type and quantity of the hostile fire expected from the enemy, and it passed through many stages in the Middle Ages. The single large tower, or keep, was replaced by a fort within a series of forts; the walls were increased in height and galleries were added for archers. The coming of gunpowder and cannon in the fifteenth century raised many new problems, because artillery fire could be concentrated in one spot, battering down weak masonry. The castle therefore lost its defensive capacity and was replaced by the bastioned fortress.

The great military engineers of the time developed the 'star' fortress, with five or six bastions each shaped in plan like the ace of spades, so that the defenders' artillery could ward off the assault forces and at the same time give covering fire to adjacent bastions. However, as we saw earlier in the Introduction, the dominance of this form was threatened in the seventeenth century by systematic methods of attack developed by Vauban. He developed fortress designs to withstand the new attack methods, and thus became the outstanding figure in both siegecraft and in fortress defence of his time. It has been said that a fortress built by Vauban was impregnable, and that one besieged by him was doomed. His contribution has been discussed by Harris [9.1].

By the early years of the twentieth century concrete began to be used in increasing quantities, and concrete roofs of 2 or 3 metres thick were built over revetments. Major cities in Europe were ringed with fortresses and in the years leading to the outbreak of the First World War the ideas of the Frenchman Brialmont were used in the construction of defensive systems. He used steel armour to protect guns, and devised retractable cupolas and turrets; his ideas were used by the French in developing their fortress system along the German frontier.

Reinforced concrete now became the standard material for protective structures, and was used in conjunction with the application of ballistics to the design of fortifications. Individual forts were often triangular in plan, and by 1914 the Brialmont system had been used to build this type of fort at Antwerp, Namur and Liege in Belgium. At the beginning of the First World War the Germans attacked through Belgium, using the Krupp 420 mm Howitzer. Liege fell in eleven days, Namur in four and Antwerp in ten, and apart from the success of the fortress of Verdun as a very strong point in the battle of the Marne, it was generally concluded at the end of the war that the cost of the fortifications was not justified.

In spite of this, as the Second World War approached, the French built the celebrated but ill-starred Maginot line, the Germans built the Siegfried line, and the Russians the Stalin line. On the Dutch/Belgian border a fort very similar to the Maginot line design was built to dominate the River Meuse. It was thought to be impregnable. On 10 May 1940 this fortress was apparently taken in a few hours by less than one hundred German glider-borne engineers. They are said to have landed directly on top of the fortifications and used new shaped hollow charges to blast into the cupolas, although the accuracy of this report has been questioned. After the breakthrough the Maginot line was outflanked, and the huge construction of mutually supporting positions stretching from Luxembourg to Switzerland, which involved 100 kilometres of tunnels, 12 million cubic metres of earthworks, 1.5 million cubic metres of concrete and 150 000 tonnes of steel, was of no value. The fortifications have been described by Taylor [9.2].

In the period since the Second World War there has been a change in the protective structure requirements of military operations, brought about by the development over the past forty years of missile systems, satellite and electronic surveillance, nuclear weapons, and sophisticated command, control and communication centres. The structures are now required to protect personnel and instrumentation against the effects of nuclear, non-nuclear and chemical weapons, to form the protective shells of weapon silos and anti-missile installations and the protective shelters of aircraft and naval craft. There has been a notable increase in experimental and theoretical studies to examine the behaviour of hardened structures under shock and impact loading, and to investigate the penetrative characteristics of high-speed missiles into metal and concrete. Attempts have also been made to formulate design rules and codes of practice for the development of this type of structure.

Much experimental data has been assembled from all the above sources, gained over a period of over 100 years, from two world wars, countless minor armed conflicts or terrorist actions, cold war defence research and weapon development programmes. A complete review of all data could fill several books, but as far as the author is aware no unified and unclassified data bank of experimental results has ever been put together.

The gradual collection of field and laboratory evidence in the early days naturally led to the development by applied mathematicians, physicists and engineering analysts of theories to predict structural behaviour, based on the fundamental criteria of dynamic loading, and requiring input loading functions, resistance and deformation relationships, mass distribution and material failure relationships. The analytical solutions for structural response, always tedious to produce, were later increased rapidly in number by the coming of the computer and by the development of theories of non-linear and limit state behaviour. Discussions of the accuracy of the empirical formulae of early design codes, so useful in making initial assessments of structural performance, have tended to be superseded by debates on the 'modelling' of material characteristics and structural actions in the all-embracing hydrocodes of modern computer analysis. The marketing of software to give instant design solutions for structures responding to explosive loading is now a business activity of some magnitude, but the assumptions on which the modelling is based are not always critically reviewed.

To weave through this forest in such a way that all trees can be inspected, if only briefly, means that a somewhat tortuous path must be followed. As an entry point we will examine the state of affairs in Britain at the beginning of the Second World War.

9.2 THEORETICAL RESPONSE

Christopherson [1.9] has reported that in 1939 the British Home Office commissioned a report on the design of buildings against air attack, the second part of which dealt with the mathematical analysis of the effect of blast on structures [9.3]. To begin with the responding structure was unrealistically treated as linear elastic and undamped, and it was assumed that it could be represented by a mass supported by a spring, capable of displacement in one direction only, and therefore having a single degree of freedom. This meant that the well-established and familiar governing equation of motion using Newton's second law could be used, and was set down in the usual form

$$\frac{m\partial^2 y}{\partial t^2} + ky = f(t),\qquad(9.1)$$

where m is the mass, k is the spring stiffness, $f(t)$ is the time-varying force, t is time and y is mass displacement.

Solutions were developed for a range of 'forcing functions', varying from a simple sinusoidal relationship between force and time to an approximately triangular relationship with an instantaneous pressure rise followed by an exponential pressure decay. This was taken as a typical force–time relationship for a detonated explosion at that time.

After the Second World War the further development of nuclear bombs resulted in intense target response research, particularly in the USA, and the basic equation (9.1) was used to provide data on the elastic analysis of simple systems in order to attempt to explain the results of the large number of response tests held on mainland USA and on small islands in the Pacific. The ratio between the displacement (y) of a system calculated from Eq. (9.1) and the displacement under the same load system applied statically (y_{st}), known as the Dynamic Load Factor, was easily shown to be related to the ratio of the duration of the applied load, t_d, and to the natural period of oscillation of the system (T). Ranges of values of y/y_{st} for a range of values of t_d/T for triangular pulses were calculated and recorded in great detail in the works of the American authors Biggs [9.4] and Newmark [9.5]

When consideration of the vibration of a one-degree of freedom elastic system after the first peak of elastic response was required, it was necessary to include damping in the analysis. On the convenient though not necessarily accurate assumption of viscous damping, Eq. (9.1) became

$$m\frac{\partial^2 y}{\partial t^2} + C\frac{\partial y}{\partial x} + ky = f(t).\qquad(9.2)$$

However, with the values of C for most structures turning out to be about one-tenth of the critical damping value ($C = 2mw$) the natural frequency of vibration was hardly changed.

It was soon clear to the research fraternity that most structures suffered permanent deformation in the face of blast loading, with the deformation governed by structural ductility as well as by elastic and inelastic properties. It was therefore necessary to consider the effect of large plastic strains and the effects of plastic energy absorption. In calculating the response of a single degree of freedom system to a suddenly applied constant load, F, for example, it was realized that there were two important and discontinuous stages. There was elastic response until deflections reached the limit of elastic behaviour, y_{el}, followed by a fully plastic response until the maximum deflection, y_m was reached. The assumption here was that the response changed from fully elastic to fully plastic with no intermediate elasto-plastic zone.

In Eqs (9.1) and (9.2) there is a relationship between (spring stiffness/mass)$^{1/2}$ and the natural circular frequency of the oscillations of the undamped spring/mass system, so that $(k/m)^2 = \omega$. Using this relationship it was shown that for the first stage of loading $y = y_{st}(1 - \text{Cos } \omega t)$ when $0 < y < y_{el}$, and for the second stage

$$y = \frac{1}{2m} (F - R_m)t_1^2 + y_{st}\omega^{t_1} \sin \omega t_{el} + y_{el}, \qquad (9.3)$$

when $y_{el} < y < y_m$.

R_m is the structural resistance when $y = y_{el}$, and $t_1 = t - t_{el}$.
The time when the maximum response was reached was given by

$$t_m = \frac{m\omega y_{st}}{R_m - F} \cdot \sin \omega t_{el}, \qquad (9.4)$$

and when this was substituted in equation (9.3) an expression was found for y_m.

Charts were produced by Biggs (9.4) and others that plotted values for y_m and t_m for ranges of values of R_m/F, and for a variety of loading functions. The triangular pulse loads were in the form of an instantaneous pressure rise followed by a linear decay, to represent detonated explosions, or a linear pressure rise followed by an equal decay, taken as an approximation to the characteristics of a dust cloud explosion. If the structural resistance was equal to or greater than the applied load, the response was entirely elastic.

The limitation of the charts was that in practice the behaviour of very few practical structures or structural components could be represented by a single degree of freedom system, but to combine multi-degree of freedom analysis with inelastic response was thought to be an almost impossible task in the days before computers. Considerable efforts were therefore made to represent a range of types of structural component by 'equivalent' single degree of freedom systems. This representation relied on the establishment of equivalent values of peak force (F_e), mass (m_e) and maximum resistance (R_{me}) so that the deflection of a significant point on the real structure was similar to that of the equivalent single degree of freedom spring/mass system. The period of initial vibration was also kept similar. This led to the idea of transformation factors, defined as $K_m = m_e/m$, $K_L = F_e/F$ and $K_R = R_{me}/R_m$, and methods were developed to calculate these for particular structural components, as described by Biggs (9.4).

It is important to note that the equivalent system was first set up on the basis of kinetic similarity, so that the maximum component displacements were accurate. Moments and shears, which depend on the derivatives of deformation patterns, rather than displacements, were less accurate. For the calculation of strains and shear forces in the component a multi-degree of freedom analysis needed to be used.

The single degree of freedom representation of structures was used in most of the design handbooks for hardened military structures produced by agencies in the USA in the period between 1960 and 1980. It was also used in the assessment of the effects of accidental explosions on civilian structures. The elasto-plastic responses of beams and plates in flexure are represented in this way in the publications of the US Army and Air Force given in references [9.6], [9.7] and [9.8], and in a related publication by Baker *et al.* [9.9]. Readers wishing to read more of the background should consult a paper by Baker and Spivey [9.10], who take the example of collapse stages in a simple beam with clamped ends under dynamic flexural loading. Elastic linear response during the early stages of loading changes to a second phase linear response after the formation of plastic hinges at the end of the beam. This phase continues until the formation of a central plastic hinge results in collapse. This 'trilinear' response and its equivalent 'bilinear' response is shown in Figure 9.1, taken from reference [9.10]. The equivalent spring constant, K_e, is $R_m/(y_{el})_e$, and the equivalent system reaches its ultimate resistance at the equivalent limit of elasticity $(y_{el})_e$. A similar representation is also made for structures such as rectangular plates with clamped edges under flexural loading, which have a four stage or 'quadrilinear' response. The transformation factors, formulae for maximum resistance and spring constants for various stages of deformation, have been established for rectangular plates using a variety of boundary

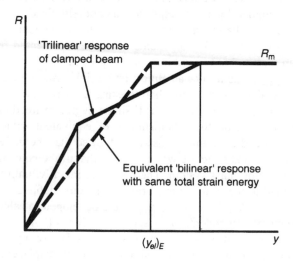

Figure 9.1 The equivalent bilinear response for a clamped beam (from Baker and Spivey, ref. 9.10).

conditions. Circular plates and other structural elements have also been treated. They appear in references [9.4] to [9.8] and in other related official publications.

The paper by Baker and Spivey [9.10] describes a computer program that solves numerically various dynamic response problems represented by the above system. The development in the late 1980s of programs like this has superseded the former tables of transformation factors. An important feature of the new programs is that dynamic reactions at supports are considered, so that dynamic shear forces can be checked. These are normally missing from the single degree of freedom spring-mass analogy. However, these dynamic reactions have been analysed on the assumption that the deflected shape of a beam, for example, remains the same after the transition from elastic to plastic response, which is not necessarily true. A very high, short-duration blast load could lead to high beam curvature close to the supports, producing high shears. Care should therefore be taken when applying the single degree of freedom analogy to explosive forces from very local or close-in detonations.

It has become customary to represent the ultimate strength of dynamically loaded structures on a pressure/impulse diagram. When the duration of the blast pressure, t_d, on an elastic structure is large compared to the natural period (T), there is very little reduction in the initial peak pressure before the structure respond elastically and reaches its maximum deflection. For values of $t_d/T > 40$, and an approximately triangular pulse, $(y/y_{st})_{max} = 2$, and deflection depends only on the peak value of the pressure and the structural stiffness. It is independent of the duration, t_d, and the mass of the structure. However, when the duration is small in comparison with the natural period, so that $t_d/T < 0.4$, deflection is proportional to impulse I, where $I = pt$. Between these extremes deflection is governed by a combination of pressure and impulse. The information for a particular structure can be represented non-dimensionally by dividing the applied pressure by that required to achieve $y/y_{st} = 2$, and by dividing the impulse by that required to produce a direct proportionality between impulse and deflection. Similar curves can be calculated for an elasto/plastic response and for other types of pressure/duration relationship.

Test results can be plotted on a pressure/impulse diagram so that the form of the transition can be established for a range of types of structure under various types of explosive attack. By these means 'iso-damage' curves are produced, which are useful to designers. As well as using test results, theoretical relationships were established in the mid-1970s by Abrahamson and Lindberg [9.11] for both elastic and rigid-plastic single degree of freedom response.

The response to explosions of complex multi-compartmentalized structures of all types seemed to demand representation by multi-degree of freedom systems, with the structural representation in the form of lumped masses connected by elastic springs, and this method was given a close examination in the 1950s. The number of separate types of motion is equal to the number of degrees of freedom, so deflections were 'coupled' and the solution involved numbers of simultaneous equations. The fundamentals of equation (9.1) still apply for undamped systems but must now be expressed in a matrix form as

$$[M]_D\{\ddot{y}\} + [K]\{y\} = \{F(t)\}, \tag{9.5}$$

where $[M]_D$ is a diagonal matrix, $\{y\}$ and $\{\ddot{y}\}$ are column matrices, $[K]$ is a stiffness matrix and $\{F(t)\}$ a column matrix of forces. When the system vibrates naturally $\{y\} = \{a_n\} \sin \omega_n t$, $\{\ddot{y}\} = -\omega_n^2\{a_n\} \sin \omega t$, and $\{F(t)\} = 0$, which gives

$$([K] - \omega_n^2[M]_D)\{a_n\} = 0.$$

Since $\{a_n\}$ is not zero, this equation can be simplified to the eigen-value relationship

$$[K] - \omega_n^2[M]_D = 0,$$

and the eigen values are equal to the squares of the natural frequencies of the modes.

The response of combinations of elements is not the only circumstance requiring multi-degree analysis. Structures buried in consolidated soil or rock create a multi-degree system involving the elastic and elasto-plastic characteristics of the earth material. It is also possible for the structure to be surrounded by earth material and then by a further layer of elemental structure such as a slab or beam grid. Methods exist for analysing this type of configuration as a multi-degree of freedom system.

Some structures that contain combinations of elements can be analysed accurately if the elements can be treated as separate single degree systems. This is possible if the kinetic energy, the internal strain energy and the work done by both systems when vibrating in a normal mode are equal. Once the equivalent system has been established, it can be shown that the dynamic load factor only depends on the applied force and angular velocity, so that the relationships developed for single-degree systems can be used. Examples were prepared by Biggs in the early 1960s and are discussed in reference [9.1]. There was a rule that if the ratio of the periods of component elements was greater than 2, then the components could be treated separately. If the ratio was less than 2, a multi-degree approach should be used.

Theoretical response methods have been augmented in recent years by the advent of non-linear explicit three-dimensional finite element codes such as DYNA-3D from the Lawrence Livermore National Laboratory in California. Calculations based on these codes have been shown to give good agreement with experimental data for a range of blast/structure problems, including structures in soil. In certain circumstances the finite element codes have been used in a preliminary way to improve the modelling used subsequently in single or multi-degree of freedom analyses. Typical examples of research in these areas are contained in papers by Bingham, Walker and Blouin [9.12], Wright and Hobbs [9.13], Terrier and Boisseau [9.14], and Crawford and Mendoza [9.15].

As these references show, comparisons between DYNA-3D and the analysis developed by Biggs have been made in conjunction with research on weapons effects on structures. But the field of offshore structures subjected to blast loads from hydrocarbon explosions has also produced comparative research. The structural response to blast of stiffened metallic panels, usually steel, involving torsional instability of the bulb-stiffeners has been examined by teams in a number of countries using DYNA-3D, ADINA, and the Biggs formulation. The use of finite element analysis has emphasized the importance of choosing correct boundary conditions. Stiffener restraint is important, as it can control the buckling mode, and it is also important to use an accurate static resistance relationship. Recent work in this area has been reported by Walker [9.16], Schleyer and Mihsein [9.17] Louca, Punjani and Harding [9.18] and Van Wees [9.19]. The non-linear response of stiffened ship panels to blast loading has been discussed by Houlston and DesRochers [9.20]. In reference [9.18] it was noted that the Biggs representation is accurate as long as a limited amount of plasticity occurs in the system. It was suggested that widespread plasticity is not reproduced by the Biggs analysis, whereas it can be accommodated in the finite element model.

9.3 LOAD FACTORS

To end our examination of loading and load effects, we must now consider load factors to be used in limit state calculations that establish the safety and reliability of structures subjected to blast or fragmentation. In establishing the factors it is first necessary to investigate how peak pressures and impulses vary when nominally similar charges are detonated. This problem has been examined in recent years in the USA, where reliability-based blast-resistant design is being actively pursued, by the recording of data from a test series in which 16 general purpose conventional bombs were exploded. It was important to establish whether the scatter of results could be adequately represented by a log-normal distribution. Reports by Carson, Morrison and Hampson [9.21] and Carson and Morrison [9.22] of tests using Mark 82 and Mark 83 General Purpose Bombs were completed in 1984 and 1987, and were examined by Twisdale, Sues, Lavelle and Miller [9.23] in 1991. In their analysis they took 325 peak pressures and 320 positive impulses at various ranges and found that the peak pressure and peak impact prediction error statistics supported the use of log-normal distributions. The incident air-blast load factors for various reliability levels are given in reference [9.23] as shown in Table 9.1.

Table 9.1

Reliability:	0.05	0.10	0.25	0.50	0.75	0.90	0.95	0.99
Load factor: (pressure)	0.65	0.72	0.86	1.05	1.28	1.54	1.71	2.09
Load factor: (impulse)	0.76	0.85	0.99	1.16	1.37	1.60	1.75	2.06

In the log-normal distribution the coefficient of variation for airblast pressure was 0.3. The authors point out that for most above-ground structures, the blast pressures that load the structure are reflected pressures, and the data for these does not compare in quality or quantity to that available for incident blast.

A later paper by Twisdale, Sues and Lavelle [9.24] indicated that the load factors in the above table should be slightly amended, because differences were noted in the mean prediction errors for half-buried bombs and bombs resting horizontally on the surface. They also reported advances in the analysis of reflected pressures.

For structures buried below ground level, a major loading is derived from ground shock, as discussed in Chapter 5, section 5.2. The variability of ground shock loads from nominally similar explosions acting through nominally similar soils is high, mainly because of the variability of soil characteristics. On the assumption that the free-field stresses were distributed on a log-normal basis, standard deviations were found from experimental results by Sues, Drake and Twisdale [9.25], using earlier stress measurements recorded by Drake *et al.* (9.26). Standard deviations from experiments for which there was proper representation of bomb and site variability showed that for soils defined as dense, wet sand, loose wet sand, and sandy clay, there was a range of standard deviation between 0.31 and 0.33 for free field velocity, and between 0.28 and 0.48 for free field stress. These figures were shown to be independent of the distance between the centre of the explosion and the target (i.e. the range).

Table 9.2 shows the relationships produced by the authors between reliability and the ground shock stress load factor for either 'typical' backfills (A) or backfills where there was much uncertainty about the soil properties (B).

Thus, for a reliability of 95%, the presence of a well-defined soil increased the load factor for pressure or free-field stress from 2.09 to 2.40, and for an uncertainly defined soil from 2.09 to 3.34. Using average soil values and taking an overall view of the accuracy of the predictions and analysis we might broadly conclude that for pressurized above ground structures and a reliability of 95%, the recommended load factor is about 2, and for buried structures under ground shock, the stress factor increases to about 3 for the same reliability level.

The team of Twisdale and Sues [9.27] have also reported on a reliability-based design procedure for penetrating fragments from an exploding bomb. The steps in this method are given by the authors in the following order: First evaluate the expected number of fragments that could be generated by the bomb, given as $W_c/2M_A^2$, where W_c is the casing weight and M_A is the fragment weight probability distribution parameter (Motts distribution). Then calculate the

Table 9.2

Reliability	0.05	0.10	0.25	0.50	0.75	0.90	0.95
Load factor (A)	0.41	0.51	0.70	1.00	1.43	1.97	2.40
Load factor (B)	0.30	0.39	0.61	1.00	1.63	2.55	3.34

expected number of fragments that could hit the structure without ricocheting using a simple spray pattern (cylindrical or spherical). The third step is to select a design fragment and find its lethality (based on penetration); using a formula for concrete as an example, lethality = $W_f^{0.4} \cdot V_s^{1.8}$, where W_f is fragment weight (lbs) and V_s the striking velocity (feet/sec). This formula is likely to be very conservative. The next steps are to calculate the probability that no fragments with a lethality greater than the design fragment will be generated by the exploding bomb, and to calculate the expected number of lethal fragments in the total fragment population. The final stages are to select a size of design fragment so that the desired level of reliability is achieved and there will be no lethal hits. This fragment is then used in the design of the protective structure. The method relies on a large amount of reliable experimental data that has been analysed on a probability basis to provide fragment weight and velocity distributions, and whether this could be assembled for a full range of weapons and circumstances is debatable.

All this research has to do with load factors, but before the overall safety level of a structure can be assessed these factors must be combined with structural resistance factors. Twisdale and his colleagues looked at this problem for loads due to projectiles and fragments, and their preliminary findings were reported in reference [9.28]. They took as an example the design of a reinforced concrete wall to resist 30 mm cannon fire with a 95% reliability, and used load factors based on existing databases for small projectiles impacting massive concrete targets. From these databases 710 records were used, of which 534 gave depth of penetration and 703 gave information on spall and perforation. A typical striking velocity coefficient of variation was 0.2, and for a reliability of 95% the specified striking velocity needed to be multiplied by a load factor (for concrete perforation) of 1.12. The factored striking velocity was then used to calculate the minimum thickness of wall to prevent perforation. For a reliability of 95% the calculated wall thickness was multiplied by a resistance factor of 1.54, so that the overall safety factor was $1.12 \times 1.54 = 1.8$. Thus the safety margin for 95% reliability in design against projectile perforation looks to be less than that for a structure under incident air blast, where just the load factor might be of the order of 2.0. These figures are influenced of course by the coefficients of variation of the experimental results, so too much should not be read into them. They should be taken as an illustration of recent thinking in the use of reliability-based analysis in the design of structures under the threat of blast or penetrating weapons.

It is interesting to compare these factors with resistance factors proposed by the author when discussing the design of underground protective structures subjected to surface air-blast loading, given in reference [9.29]. Using engineering judgement a load factor of 1.5 was proposed for dynamic pressures due to explosions on the surface, and a resistance factor of 1.6 for strength variations due to uneven soil compaction or the presence of ground water. For structures in which close attention was paid to emplacement it was considered that the resistance factor might be reduced to about 1.3, giving an overall factor of 1.95. This is close to the factors produced in the reliability based air-blast analysis described earlier.

The work on reliability-based design is still continuing, and no doubt there will be adjustments to the proposals of Twisdale and others as time goes on, particularly as more test information becomes available and as analytical ideas progress. It is encouraging to know, however, that modern load and resistance factor design has now entered the world of structural assessment under blast and penetrative loading, as it has for buildings, bridges, aircraft and ships under dynamic loads due to wind, waves and earthquakes.

9.4 EVOLUTION

We have now examined the gradual assembly of information on the explosive loading of structures over two centuries and more. As in many other subjects, the march of progress has not been gradual and well-conditioned. There have been periods of intense activity followed by years of rest and recovery, not unlike the growth patterns of nature which so often consist of steep gradients and level plateaux. From the analytical and experimental viewpoints, major progress was made at the time of two world wars and the nuclear cold war, when money was available for the very costly activity of full-scale field testing. Theories and analysis are subject to swings of fashion, but the results of well-conducted tests have a lasting value.

There are psychological problems when a field of research is connected with an activity considered to be associated with physical danger, or with the pursuit of war or terrorism. A background of uncertainty can sometimes lead to a superficial consideration of basic engineering or scientific principles. Much has been written in this book about the behaviour of engineering structures, but less about the behaviour of structural engineers. Perhaps future research will examine this problem in greater depth, and the subject has been explored by the author in a series of short stories.

Future work on the shock loading of structures is likely to continue at a steady rate as the train of research moves on to its next stop. On the way it is expected to journey through fields of risk and reliability, modelling laws, new instrumentation, and the computer simulation of large-scale testing. There is also a considerable interest now in the prediction of fragment loading and damage, and in penetration mechanics.

At the time of writing these closing remarks the 8th International Symposium on the Interaction of the Effects of Munitions with Structures has just taken place in the USA. A glance through the agenda shows that, very approximately, about half the papers dealt with structural response and protective design, about one quarter with penetration and fragmentation, and the remaining quarter was divided about equally between blast loading research and testing or simulation.

There were many papers on the application of hydrocodes and other numerical techniques to the various problem areas of the subject.

One of the main tasks of the near future will be to bring the fruits of structural loading research into a form suitable for design codes of practice for civil buildings, bridges and other land, sea and air structures. It is hoped that the research history and inspirations from the past retailed in this book will help towards this work, and at the same time define a background for the evolution of new scientific insights as well as the development of existing knowledge.

9.5 REFERENCES

9.1 Harris, A. (1989) A man of mettle, *New Civil Engineer*, 18 May.
9.2 Taylor, G. (1987) The general design and use of hardened defences in twentieth century warfare, *Royal Engineers' Journal*, **101**(2).
9.3 Anon (1939) *The Design of Buildings against Air Attack, Part 2, Mathematical Analysis of the Effect of Blast on Structures*, HOPP/18, Civil Defence Research Committee paper RC 23.
9.4 Biggs, J. M. (1964) *Introduction to Structural Dynamics*, McGraw Hill, New York.
9.5 Newmark, N. M. *et al.* (1961, rev. 1964) *Design of Structures to Resist Nuclear Weapons Effects*, American Society of Civil Engineers, Manual of Engineering Practice No. 42. Republished in 1985.
9.6 US Dept of the Army (1969 and 1991) *Structures to Resist the Effects of Accidental Explosions*, US Dept of the Army Technical Manual TM5-1300.
9.7 Crawford, R. E., Higgins, C. J.and Bultmann, E. H. (1974) *The Air Force Manual for Design and Analysis of Hardened Structures*, AFWL TR 74-102, AFWL Kirkland AFB, New Mexico, October.
9.8 US Dept of the Army (1986) *Fundamentals of Protective Design for Conventional Weapons*, US Army TM5-855-1, November.
9.9 Baker, W. E. *et al.* (1983) *Explosion Hazards and Evaluation*, Elsevier, New York.
9.10 Baker, W E., and Spivey, K H. (1989) BIGGS – simplified elastic-plastic dynamic response. In P. S. Bulson (ed.), *Structures under Shock and Impact*, Elsevier, p. 135.
9.11 Abrahamson, G. R. and Lindberg, H. E. (1976) Peak load – Impulse characterisation of critical pulse loads in structural dynamics, *Nuclear Engineering and Design*, 37, pp. 35–46.
9.12 Bingham, B. L., Walker, R. E. and Blouin, S. E. (1993) Response of pile foundations in saturated soil, *Proc. 6th Int. Symp. on Interaction of Non-nuclear Munitions with Structures*, Panama City Beach, Florida, USA, May.
9.13 Wright, S. J. and Hobbs, B. (1994) Use of the DYNA-3D FE code to quantify the response of steel plate girders subjected to localised combined blast and fragment loads. In P. S. Bulson (ed.), *Structures under Shock and Impact*, Computational Mechanics Publications, June.
9.14 Terrier, J. M. and Boisseau, J. F. X. (1989) Numerical simulations of reinforced structure response subjected to high explosive detonation, *Proc. 4th Int. Symp. on Interaction of Non-nuclear Munitions with Structures*, Panama City Beach, Florida, USA, April.
9.15 Crawford, J. E. and Mendoza, P. J. (1985) Combined finite element and lumped mass techniques for parametric structural analysis of structures, *Proc. 2nd Symp. on the Interaction of Non-nuclear Munitions with Structures*, Panama City Beach, Florida, USA, April.

9.16 Walker, A. C. (1993) Non-linear analysis of blast walls to evaluate safe ultimate loads, *2nd Int. Conf. on Offshore Structural Design against Extreme Loads*, ERA, London.
9.17 Schleyer, G. and Mihsein, M. (1992) Development of mathematical models for dynamic analysis of structures, *1st Int. Conf. on Structural Design against Extreme Loads*, ERA, London.
9.18 Louca, L. A., Punjani, M. and Harding, J. E. (1996) Non-linear analysis of blast walls and stiffened panels subjected to hydrocarbon explosions, *Journal of Constructional Steel Research*, **37**(2), April.
9.19 Van Wees, R. M. (1993) Design of stiffened panels to withstand explosion loads, *2nd Int. Conf. on Offshore Structural Design against Extreme Loads*, ERA, London.
9.20 Houlston, R. and DesRochers, C. G. (1987) Non-nuclear structural response of ship panels subjected to air blast loading, *Proc. 6th Conf. on Non-linear Analysis*, ADINA.
9.21 Carson, J. M., Morrison, D. and Hampson, R. J. (1984) *Conventional High Explosive Blast and Shock (CHEBS) Test Series: Mark 82 General Purpose Bomb Tests*, Air Force Weapons Laboratory, AFNL-TR-84-27, June.
9.22 Carson, J. M. and Morrison D. (1987) *Conventional High Explosive Blast and Shock (CHEBS) Test Series: Mark 83 General Purpose Bomb Tests*, Air Force Weapons Laboratory, AFWL-TR-86-53, Parts 1–3, January.
9.23 Twisdale, L. A., Sues, R. H., Lavelle, F. M. and Miller, D. B. (1991) Research to develop reliability-based design methodology for protective structures, *Proc. 5th Int. Symp. on the Interaction of Conventional Munitions with Protective Structures*, Mannheim, Germany, April.
9.24 Twisdale, L. A., Sues, R. H. and Lavelle, F. M. (1993) Reliability based design methods for the Protective Construction Design Manual, *Proc. 6th Int. Symp. on the Interaction of Non-nuclear Munitions with Structures*, Panama City Beach, Florida, May.
9.25 Sues, R. H., Drake, S. L., and Twisdale, L. A. (1993) Reliability based safety factors for grounds shock loads in protective construction, *Proc. 6th Int. Symp. on the Interaction of Non-nuclear Munitions with Structures*, Panama City Beach, Florida, May.
9.26 Drake, J. L. *et al.* (1989) *Protective Construction Design Manual*, ESL-TR-87-57, Air Force Engineering and Services Centre, Tyndall Air Force Base, November.
9.27 Sues, R. H. and Twisdale, L. A. (1993) How to select a design fragment for protective structure design with consistent reliability, *Proc. 6th Int. Symp. on the Interaction of Non-nuclear Munitions with Structures*, Panama City Beach, Florida, May.
9.28 Sues, R. H., Hwang, C. W., Twisdale, L. A. and Lavelle, F. M. (1991) Reliability based design of R/C structures for protection against projectiles and fragments, *Proc. 5th Int. Symp. on the Interaction of Conventional Munitions with Protective Structures*, Mannheim, Germany, April.

Author index

Subject index